Introduction to Petrology

Introduction to Petrology

George Hamill

Larsen & Keller
www.larsen-keller.com

Introduction to Petrology
George Hamill
ISBN: 978-1-64172-641-2 (Hardback)

Larsen & Keller
Published by Larsen and Keller Education,
5 Penn Plaza,
19th Floor,
New York, NY 10001, USA

Cataloging-in-Publication Data

Introduction to petrology / George Hamill.
p. cm.
Includes bibliographical references and index.
ISBN 978-1-64172-641-2
1. Petrology. 2. Rocks. 3. Physical geology. I. Hamill, George.
QE431.2. I58 2022
552--dc23

For more information regarding Larsen and Keller Education and its products, please visit the publisher's website www.larsen-keller.com

Table of Contents

Preface

The study of rocks along with the processes through which they are formed is conducted under the branch of geology known as petrology. There are three major subdivisions within petrology, namely, sedimentary, metamorphic and igneous petrology. Sedimentary petrology deals with the texture and composition of sedimentary rocks. Igneous petrology is the study of the composition and texture of igneous rocks such as volcanic and plutonic rocks. Metamorphic petrology deals with the rocks which were initially sedimentary or igneous but due to extreme pressure or temperature, have undergone chemical or mineralogical changes. Some of the fields used within petrology are mineralogy, petrography and chemical analysis. This book provides comprehensive insights into the field of petrology. It consists of contributions made by international experts. Coherent flow of topics, student-friendly language and extensive use of examples make this book an invaluable source of knowledge.

To facilitate a deeper understanding of the contents of this book a short introduction of every chapter is written below:

Chapter 1- The branch of geology which studies rocks as well as the conditions under which they are formed is known as petrology. There are three subdivisions in petrology, namely, igneous, metamorphic and sedimentary petrology. This is an introductory chapter which will introduce briefly all the significant aspects of the different subdivisions of petrology.

Chapter 2- The rocks which are formed through the cooling and solidification of magma or lava are known as igneous rocks. They occur in a wide range of geological locations such as platforms, basins and extended crust and oceanic crust. A few examples of igneous rocks are obsidian, diorite, andesite and diabase. The topics elaborated in this chapter will help in gaining a better perspective about the different kinds of igneous rocks.

Chapter 3- The rocks which are formed by the accumulation of small particles and subsequent cementation of mineral or organic particles on the floor of bodies of water at the Earth's surface are known as sedimentary rocks. This chapter has been carefully written to provide an easy understanding of the varied types of sedimentary rocks such as breccia, conglomerate and siltstone.

Chapter 4- The rocks which arise from the transformation of existing rock types through a process called metamorphism are known as metamorphic rocks. Some of the different types of metamorphism are contact or thermal metamorphism, regional metamorphism, hydrothermal metamorphism and plutonic metamorphism. The chapter closely examines these key types of metamorphic rocks to provide an extensive understanding of the subject.

Chapter 5- The description of a rock unit in terms of its physical characteristics visible at outcrop, in hand or core samples, or by using low magnification microscopy is known as lithology. The topics elaborated in this chapter will help in gaining a better perspective about the discipline of lithology as well as related disciplines of lithostratigraphy and stratigraphy.

I owe the completion of this book to the never-ending support of my family, who supported me throughout the project.

George Hamill

1

Understanding Petrology

The branch of geology which studies rocks as well as the conditions under which they are formed is known as petrology. There are three subdivisions in petrology, namely, igneous, metamorphic and sedimentary petrology. This is an introductory chapter which will introduce briefly all the significant aspects of the different subdivisions of petrology.

Rock

Rock in geology means naturally occurring and coherent aggregate of one or more minerals. Such aggregates constitute the basic unit of which the solid Earth is comprised and typically form recognizable and mappable volumes. Rocks are commonly divided into three major classes according to the processes that resulted in their formation. These classes are (1) igneous rocks, which have solidified from molten material called magma; (2) sedimentary rocks, those consisting of fragments derived from preexisting rocks or of materials precipitated from solutions; and (3) metamorphic rocks, which have been derived from either igneous or sedimentary rocks under conditions that caused changes in mineralogical composition, texture, and internal structure. These three classes, in turn, are subdivided into numerous groups and types on the basis of various factors, the most important of which are chemical, mineralogical, and textural attributes.

General Considerations

Rock Cycle

Geologic materials—mineral crystals and their host rock types—are cycled through various forms. The process depends on temperature, pressure, time, and changes in environmental conditions in the Earth's crust and at its surface. The rock cycle illustrated in figure below reflects the basic relationships among igneous, metamorphic, and sedimentary rocks. Erosion includes weathering (the physical and chemical breakdown of minerals) and transportation to a site of deposition. Diagenesis is the process of forming sedimentary rock by compaction and natural cementation of grains, or crystallization from water or solutions, or recrystallization. The conversion of sediment to rock is termed lithification.

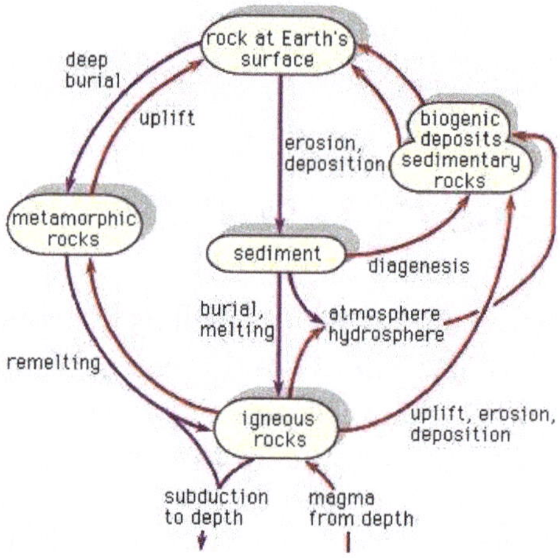

Figure: The rock cycle.

Texture

The texture of a rock is the size, shape, and arrangement of the grains (for sedimentary rocks) or crystals (for igneous and metamorphic rocks). Also of importance are the rock's extent of homogeneity (i.e., uniformity of composition throughout) and the degree of isotropy. The latter is the extent to which the bulk structure and composition are the same in all directions in the rock.

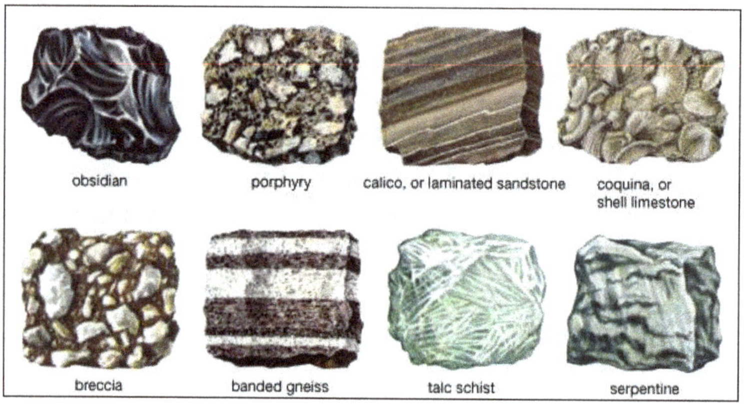

As depicted in these above figure. Rocks have many different textures. Layered sandstone produces a gritty texture, whereas coquina may be rough with cemented shells occasionally producing a sharp edge. Likewise, breccia, which contains pieces of other rocks that have been cemented together, and porphyry, which contains interlocking mineral crystals, tend to be rough. In contrast, obsidian tends to have a smooth glassy feel, whereas serpentine may feel platy or fibrous, and talc schist often feels greasy. On the other hand, the texture of gneiss is often described by its distinct banding.

Analysis of texture can yield information about the rock's source material, conditions and environment of deposition (for sedimentary rock) or crystallization and recrystallization (for igneous and metamorphic rock, respectively), and subsequent geologic history and change.

Classification by Grain or Crystal Size

The common textural terms used for rock types with respect to the size of the grains or crystals. The particle-size categories are derived from the Udden-Wentworth scale developed for sediment. For igneous and metamorphic rocks, the terms are generally used as modifiers—e.g., medium-grained granite. Aphanitic is a descriptive term for small crystals, and phaneritic for larger ones. Very coarse crystals (those larger than 3 centimetres, or 1.2 inches) are termed pegmatitic.

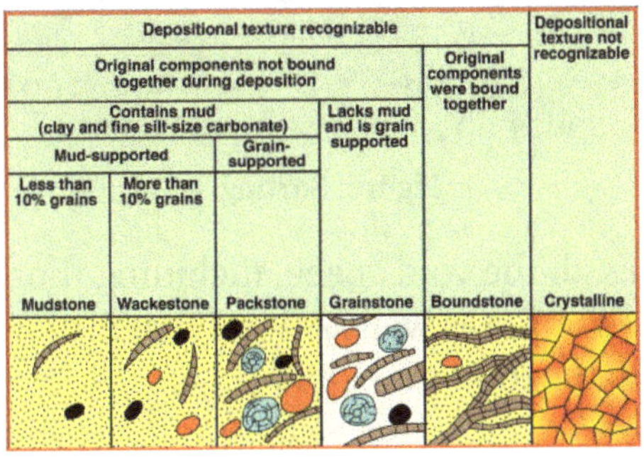

For sedimentary rocks, the broad categories of sediment size are coarse (greater than 2 millimetres, or 0.08 inch), medium (between 2 and 1/16 millimetres), and fine (under 1/16 millimetre). The latter includes silt and clay, which both have a size indistinguishable by the human eye and are also termed dust. Most shales (the lithified version of clay) contain some silt. Pyroclastic rocks are those formed from clastic material ejected from volcanoes. Blocks are fragments broken from solid rock, while bombs are molten when ejected.

Porosity

The term rock refers to the bulk volume of the material, including the grains or crystals as well as the contained void space. The volumetric portion of bulk rock that is not occupied by grains, crystals, or natural cementing material is termed porosity. That is to say, porosity is the ratio of void volume to the bulk volume (grains plus void space). This void space consists of pore space between grains or crystals, in addition to crack space. In sedimentary rocks, the amount of pore space depends on the degree of compaction of the sediment (with compaction generally increasing with depth of burial), on the packing arrangement and shape of grains, on the amount of cementation, and on the degree of sorting. Typical cements are siliceous, calcareous or carbonate, or iron-bearing minerals.

Sorting is the tendency of sedimentary rocks to have grains that are similarly sized—i.e., to have a narrow range of sizes. Poorly sorted sediment displays a wide range of grain sizes and hence has decreased porosity. Well-sorted indicates a grain size distribution that is fairly uniform. Depending on the type of close-packing of the grains, porosity can be substantial. It should be noted that in engineering usage—e.g., geotechnical or civil engineering—the terminology is phrased oppositely and is referred to as grading. A well-graded sediment is a (geologically) poorly sorted one, and a poorly graded sediment is a well-sorted one.

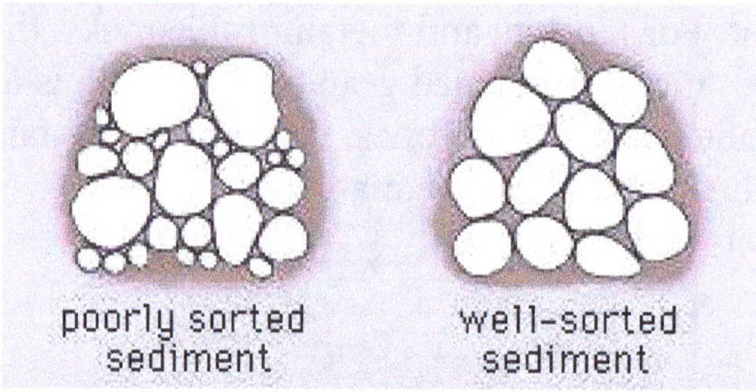

Figure: Sorting.

Total porosity encompasses all the void space, including those pores that are interconnected to the surface of the sample as well as those that are sealed off by natural cement or other obstructions. Thus the total porosity (ϕ_T) is

$$\varphi r = 100\left(1 - \frac{Vol_G}{Vol_B}\right)\%.$$

Where Vol_G is the volume of grains (and cement, if any) and Vol_B is the total bulk volume. Alternatively, one can calculate ϕ_T from the measured densities of the bulk rock and of the (mono)mineralic constituent. Thus,

$$\varphi r = 100\left(1 - \frac{\rho_B}{\rho_G}\right)\%,$$

where ρB is the density of the bulk rock and ρG is the density of the grains (*i.e.*, the mineral, if the composition is monomineralogic and homogeneous). For example, if a sandstone has a ρ_B of 2.38 grams per cubic centimetre (g/cm³) and is composed of quartz (SiO_2) grains having ρ_G of 2.65 g/cm³, the total porosity is

$$\varphi r = 100\left(1 - \frac{2.38}{2.65}\right) = 10.2\%.$$

Apparent (effective, or net) porosity is the proportion of void space that excludes the sealed-off pores. It thus measures the pore volume that is effectively interconnected and accessible to the surface of the sample, which is important when considering the storage and movement of subsurface fluids such as petroleum, groundwater, or contaminated fluids.

Physical Properties

Physical properties of rocks are of interest and utility in many fields of work, including geology, petrophysics, geophysics, materials science, geochemistry, and geotechnical engineering. The scale of investigation ranges from the molecular and crystalline up to terrestrial studies of the Earth and other planetary bodies. Geologists are interested in the radioactive age dating of rocks to reconstruct the origin of mineral deposits; seismologists formulate prospective earthquake predictions using premonitory physical or chemical changes; crystallographers study the synthesis of minerals with special optical or physical properties; exploration geophysicists investigate the variation of physical properties of subsurface rocks to make possible detection of natural resources such as oil and gas, geothermal energy, and ores of metals; geotechnical engineers examine the nature and behaviour of the materials on, in, or of which such structures as buildings, dams, tunnels, bridges, and underground storage vaults are to be constructed; solid-state physicists study the magnetic, electrical, and mechanical properties of materials for electronic devices, computer components, or high-performance ceramics; and petroleum reservoir engineers analyze the response measured on well logs or in the processes of deep drilling at elevated temperature and pressure.

Since rocks are aggregates of mineral grains or crystals, their properties are determined in large part by the properties of their various constituent minerals. In a rock these general properties are determined by averaging the relative properties and sometimes orientations of the various grains or crystals. As a result, some properties that are anisotropic (i.e., differ with direction) on a submicroscopic or crystalline scale are fairly isotropic for a large bulk volume of the rock. Many properties are also dependent on grain or crystal size, shape, and packing arrangement, the amount and distribution of void space, the presence of natural cements in sedimentary rocks, the temperature and pressure, and the type and amount of contained fluids (e.g., water, petroleum, gases). Because many rocks exhibit a considerable range in these factors, the assignment of representative values for a particular property is often done using a statistical variation.

Some properties can vary considerably, depending on whether measured in situ (in place in the subsurface) or in the laboratory under simulated conditions. Electrical resistivity, for example, is highly dependent on the fluid content of the rock in situ and the temperature condition at the particular depth.

Density

Density varies significantly among different rock types because of differences in mineralogy and porosity. Knowledge of the distribution of underground rock densities can assist in interpreting subsurface geologic structure and rock type.

In strict usage, density is defined as the mass of a substance per unit volume; however, in common usage, it is taken to be the weight in air of a unit volume of a sample at a specific temperature. Weight is the force that gravitation exerts on a body (and thus varies with location), whereas mass (a measure of the matter in a body) is a fundamental property and is constant regardless of location. In routine density measurements of rocks, the sample weights are considered to be equivalent to their masses, because the discrepancy between weight and mass would result in less error on the computed density than the experimental errors introduced in the measurement of volume. Thus, density is often determined using weight rather than mass. Density should properly be reported in kilograms per cubic metre (kg/m³), but is still often given in grams per cubic centimetre (g/cm³).

Another property closely related to density is specific gravity. It is defined, as noted above, as the ratio of the weight or mass in air of a unit volume of material at a stated temperature to the weight or mass in air of a unit volume of distilled water at the same temperature. Specific gravity is dimensionless (*i.e.*, has no units).

The bulk density of a rock is $\rho_B = W_G/V_B$, where W_G is the weight of grains (sedimentary rocks) or crystals (igneous and metamorphic rocks) and natural cements, if any, and *VB* is the total volume of the grains or crystals plus the void (pore) space. The density can be dry if the pore space is empty, or it can be saturated if the pores are filled with fluid (*e.g.*, water), which is more typical of the subsurface (in situ) situation. If there is pore fluid present,

$$(\text{saturated})\,\rho_B = \frac{W_G + W_{fl}}{V_B},$$

where W_{fl} is the weight of pore fluid. In terms of total porosity, saturated density is

$$(\text{saturated})\,\rho_B = \rho_G\left(1 - \varphi r\right) + \rho_{fl\varphi r},$$

and thus,

$$\varphi r = \frac{\rho_G - \rho_B}{\rho_G - \rho_{fl}},$$

where ρ_{fl} is the density of the pore fluid. Density measurements for a given specimen involve the determination of any two of the following quantities: pore volume, bulk volume, or grain volume, along with the weight.

A useful way to assess the density of rocks is to make a histogram plot of the statistical range of a set of data. The representative value and its variation can be expressed as follows: (1) mean, the average value, (2) mode, the most common value (i.e., the peak of the distribution curve), (3) median, the value of the middle sample of the data set (i.e., the value at which half of the samples are below and half are above), and (4) standard deviation, a statistical measure of the spread of the data (plus and minus one standard deviation from the mean value includes about two-thirds of the data).

A compilation of dry bulk densities for various rock types found in the upper crust of the Earth is listed in the table below. A histogram plot of these data, giving the percent of the samples as a function of density is shown in figure below. The parameters given include sample division, the range of density in one data column—*e.g.*, 0.036 g/cm³ for figure below, number of samples, and standard deviation. The small inset plot is the percentage of samples (on the vertical axis) that lie within the interval of the "mode - *x*" to the "mode + *x*," where *x* is the horizontal axis.

Table: Dry bulk densities for various rock types.

Rock type	Number of samples	Mean (grams per cubic cm)	Standard Deviation	Mode (grams per cubic cm)	Median (grams per cubic cm)
All rocks	1,647	2.73	0.26	2.65	2.86
Andesite	197	2.65	0.13	2.58	2.66
Basalt	323	2.74	0.47	2.88	2.87
Diorite	68	2.86	0.12	2.89	2.87
Dolerite (diabase)	224	2.89	0.13	2.96	2.90
Gabbro	98	2.95	0.14	2.99	2.97
Granite	334	2.66	0.06	2.66	2.66
Quartz porphyry	76	2.62	0.06	2.60	2.62
Rhyolite	94	2.51	0.13	2.60	2.49
Syenite	93	2.70	0.10	2.67	2.68
Trachyte	71	2.57	0.10	2.62	2.57
Sandstone	107	2.22	0.23	2.22	2.22

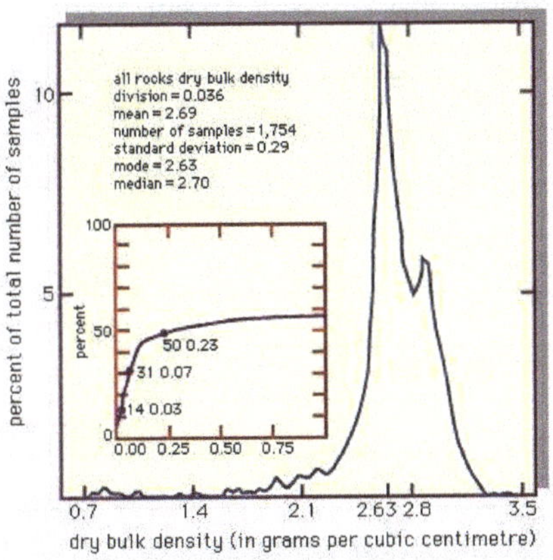

Figure: Dry bulk densities (distribution with density) for all rocks given in table above.

In the figure above, the most common (modal) value of the distribution falls at 2.63 g/cm³, roughly the density of quartz, an abundant rock-forming mineral. Few density values for these upper crustal rocks lie above 3.3 g/cm³. A few fall well below the mode, even occasionally under 1 g/cm³. The reason for this is shown in figure above, which illustrates the density distributions for granite, basalt, and sandstone. Granite is an intrusive igneous rock with low porosity and a well-defined chemical (mineral) composition; its range of densities is narrow. Basalt is, in most cases, an extrusive igneous rock that can exhibit a large variation in porosity (because entrained gases leave voids called vesicles), and thus some highly porous samples can have low densities. Sandstone is a clastic sedimentary rock that can have a wide range of porosities depending on the degree of sorting, compaction, packing arrangement of grains, and cementation. The bulk density varies accordingly.

Other distribution plots of dry bulk densities are given in table below, with a sample division of 0.036 g/cm³ for Figures and and of 0.828 percent for Figure. The Table lists typical ranges of dry bulk densities for a variety of other rock types as prepared by the American geologists Gordon R. Johnson and Gary R. Olhoeft.

Typical density ranges for some other rock types	
Rock type	Density (grams per cubic cm)
Amphibolite	2.79–3.14
Andesite glass	2.40–2.57
Anhydrite	2.82–2.93
Anorthosite	2.64–2.92
Basalt glass	2.70–2.85
Chalk	2.23
Dolomite	2.72–2.84
Dunite	2.98–3.76
Eclogite	3.32–3.45
Gneiss	2.59–2.84
Granodiorite	2.67–2.78
Limestone	1.55–2.75
Marble	2.67–2.75
Norite	2.72–3.02
Peridotite	3.15–3.28
Quartzite	2.65
rock salt	2.10–2.20
schist	2.73–3.19
shale	2.06–2.67
slate	2.72–2.84

The density of clastic sedimentary rocks increases as the rocks are progressively buried. This is because of the increase of overburden pressure, which causes compaction, and the progressive cementation with age. Both compaction and cementation decrease the porosity.

Mechanical Properties

Stress and Strain

When a stress σ (force per unit area) is applied to a material such as rock, the material experiences a change in dimension, volume, or shape. This change, or deformation, is called strain (ε). Stresses can be axial—e.g., directional tension or simple compression—or shear (tangential), or all-sided (e.g., hydrostatic compression). The terms stress and pressure are sometimes used interchangeably, but often stress refers to directional stress or shear stress and pressure (P) refers to hydrostatic compression. For small stresses, the strain is elastic (recoverable when the stress is removed and linearly proportional to the applied stress). For larger stresses and other conditions, the strain can be inelastic, or permanent.

Elastic Constants

In elastic deformation, there are various constants that relate the magnitude of the strain response to the applied stress. These elastic constants include the following:

1. Young's modulus (E) is the ratio of the applied stress to the fractional extension (or shortening) of the sample length parallel to the tension (or compression). The strain is the linear change in dimension divided by the original length.
2. Shear modulus (μ) is the ratio of the applied stress to the distortion (rotation) of a plane originally perpendicular to the applied shear stress; it is also termed the modulus of rigidity.
3. Bulk modulus (k) is the ratio of the confining pressure to the fractional reduction of volume in response to the applied hydrostatic pressure. The volume strain is the change in volume of the sample divided by the original volume. Bulk modulus is also termed the modulus of incompressibility.
4. Poisson's ratio (σ_p) is the ratio of lateral strain (perpendicular to an applied stress) to the longitudinal strain (parallel to applied stress).

For elastic and isotropic materials, the elastic constants are interrelated. For example,

$$\sigma_p = \frac{E}{2\mu} - 1:$$

$$E = 3k\left(1 - 2\sigma_p\right);$$

And

$$\mu = \frac{3}{2}k\left(\frac{1 - 2\sigma_p}{1 + \sigma_p}\right).$$

The following are the common units of stress:

1 bar = 10^6 dynes per square centimetre

= 10^5 newtons per square metre, or pascals (Pa)

= 0.1 megapascal (i.e., 0.1×10^6 pa).

Thus 10 kilobars = 1 gigapascal (*i.e.*, 10^9 Pa).

Rock Mechanics

The study of deformation resulting from the strain of rocks in response to stresses is called rock mechanics. When the scale of the deformation is extended to large geologic structures in the crust of the Earth, the field of study is known as geotectonics.

The mechanisms and character of the deformation of rocks and Earth materials can be investigated through laboratory experiments, development of theoretical models based on the properties of materials, and study of deformed rocks and structures in the field. In the laboratory, one can simulate—either directly or by appropriate scaling of experimental parameters—several conditions. Two types of pressure may be simulated: confining (hydrostatic), due to burial under rock overburden, and internal (pore), due to pressure exerted by pore fluids contained in void space in the rock. Directed applied stress, such as compression, tension, and shear, is studied, as are the effects of increased temperature introduced with depth in the Earth's crust. The effects of the duration of time and the rate of applying stress (i.e., loading) as a function of time are examined. Also, the role of fluids, particularly if they are chemically active, is investigated.

Some simple apparatuses for deforming rocks are designed for biaxial stress application: a directed (uniaxial) compression is applied while a confining pressure is exerted (by pressurized fluid) around the cylindrical specimen. This simulates deformation at depth within the Earth. An independent internal pore-fluid pressure also can be exerted. The rock specimen can be jacketed with a thin, impermeable sleeve (e.g., rubber or copper) to separate the external pressure medium from the internal pore fluids. The specimen is typically a few centimetres in dimension.

Another apparatus for exerting high pressure on a sample was designed in 1968 by Akira Sawaoka, Naoto Kawai, and Robert Carmichael to give hydrostatic confining pressures up to 12 kilobars (1.2 gigapascal), additional directed stress, and temperatures up to a few hundred degrees Celsius. The specimen is positioned on the baseplate; the pressure is applied by driving in pistons with a hydraulic press. The end caps can be locked down to hold the pressure for time experiments and to make the device portable.

Apparatuses have been developed, typically using multianvil designs, which extend the range of static experimental conditions—at least for small specimens and limited times—to pressures as high as about 1,700 kilobars and temperatures of about 2,000 °C. Such work has been pioneered by researchers such as Peter M. Bell and Ho-Kwang Mao, who conducted studies at the Geophysical Laboratory of the Carnegie Institution in Washington, D.C. Using dynamic techniques (i.e., shock from explosive impact generated by gun-type designs), even higher pressures up to 7,000 kilobars (700 gigapascal)—which is nearly twice the pressure at the centre of the Earth and seven million times greater than the atmospheric pressure at the Earth's surface—can be produced for very short times. A leading figure in such ultrapressure work is A. Sawaoka at the Tokyo Institute of Technology.

In the upper crust of the Earth, hydrostatic pressure increases at the rate of about 320 bars per kilometre, and temperature increases at a typical rate of 20°–40 °C per kilometre, depending on recent crustal geologic history. Additional directed stress, as can be generated by large-scale crustal deformation (tectonism), can range up to 1 to 2 kilobars. This is approximately equal to the ultimate strength (before fracture) of solid crystalline rock at surface temperature and pressure. The stress released in a single major earthquake—a shift on a fault plane—is about 50–150 bars.

In studying the deformation of rocks one can start with the assumption of ideal behaviour: elastic strain and homogeneous and isotropic stress and strain. In reality, on a microscopic scale there are grains and pores in sediments and a fabric of crystals in igneous and metamorphic rocks. On a large scale, rock bodies exhibit physical and chemical variations and structural features. Furthermore, conditions such as extended length of time, confining pressure, and subsurface fluids affect the rates of change of deformation. The figure below shows the generalized transition from brittle fracture through faulting to plastic-flow deformation in response to applied compressional stress and the progressive increase of confining pressure.

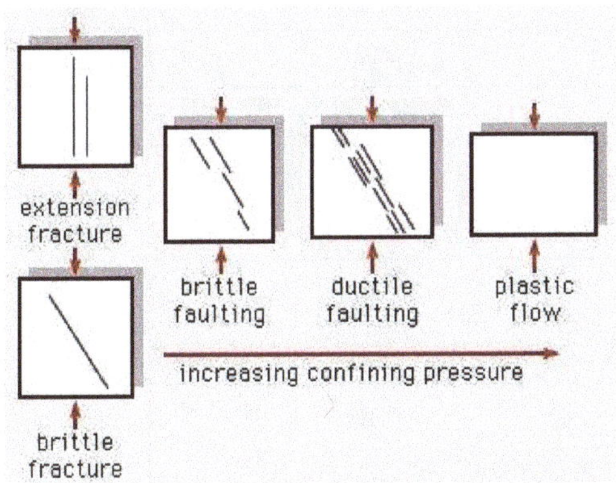

Figure: Deformation as affected by increased confining pressure

Stress-strain Relationships

The deformation of materials is characterized by stress-strain relations. For elastic-behaviour materials, the strain is proportional to the load (i.e., the applied stress). The strain is immediate with stress and is reversible (recoverable) up to the yield point stress, beyond which permanent strain results. For viscous material, there is laminar (slow, smooth, parallel) flow; one must exert a force to maintain motion because of internal frictional resistance to flow, called the viscosity. Viscosity varies with the applied stress, strain rate, and temperature. In plastic behaviour, the material strains continuously (but still has strength) after the yield point stress is reached; however, beyond this point there is some permanent deformation. In elasticoviscous deformation, there is combined elastic and viscous behaviour. The material yields continuously (viscously) for a constant applied load. An example of such behaviour is creep, a slow, permanent, and continuous deformation occurring under constant load over a long time in such materials as crystals, ice, soil and sediment, and rocks at depth. In firmoviscous behaviour, the material is essentially solid but the strain is not immediate with application of stress; rather, it is taken up and released exponentially. A plasticoviscous material exhibits elastic behaviour for initial stress (as in plastic behaviour), but after the yield point stress is reached, it flows like a viscous fluid.

Some representative values of elastic constants and properties are listed in table below. The coefficient of viscosity (η) is the ratio of applied stress to the rate of straining (change of strain with time). It is measured in units of poise; one poise equals one dyne-second per square centimetre.

Some typical values of elastic constants and properties		
	Elastic constants (at room temperature and pressure)	
Material	Young's modulus (in 10^6 bars)	Shear modulus (in 10^6 bars)
Ice	0.1	0.03
Shale	0.2–0.3	0.15
Limestone	0.4–0.7	0.22–0.26
Granite	0.3–0.6	0.2
Basalt	0.7–0.9	0.3
Steel	2.1	0.83
Material	Temperature (degrees Celsius)	Coefficient of viscosity (poises)
Lava (Mount Vesuvius)	1,100	28,300
	1,400	250
Lava (Oshima, Japan)	1,038	230,000
	1,125	5,600
Andesite lava	1,400	150–1,500
Material	Compressive strength (at room temperature and pressure, in kilobars)	
Shale	0.8–1.8	
Sandstone	0.5–2	
Limestone	1–2	
Granite	1.7–2.5	
Basalt	1–3.4	

Rheology is the study of the flow deformation of materials. The concept of rheidity refers to the capacity of a material to flow, arbitrarily defined as the time required with a shear stress applied for the viscous strain to be 1,000 times greater than the elastic strain. It is thus a measure of the threshold of fluidlike behaviour. Although such behaviour depends on temperature, relative comparisons can be made. Some representative values of rheidity times are given in the table.

Rheidity threshold of fluidlike deformation	
Material	Approximate time
Ice (e.g., glacier)	2 weeks
Gypsum	1 year
Rock salt (e.g., saltdome)	10–20 years
Serpentine (a mafic silicate mineral)	10,000 years

Typical stress-strain (deformation) curves for rock materials are shown in the figure below. The stress σ, compression in the figure below, is force per unit area. The strain ε is fractional shortening of the specimen parallel to the applied compression; it is given here in percent. The brittle material behaves elastically nearly until the point of fracture (denoted X), whereas the ductile (plastically deformable) material is elastic up to the yield point but then has a range of plastic deformation before fracturing. The ability to undergo large permanent deformation before fracture is called ductility. For plastic deformation, the flow mechanisms are intracrystalline (slip and twinning within crystal grains), intercrystalline motion by crushing and fracture (cataclasis), and recrystallization by solutioning or solid diffusion.

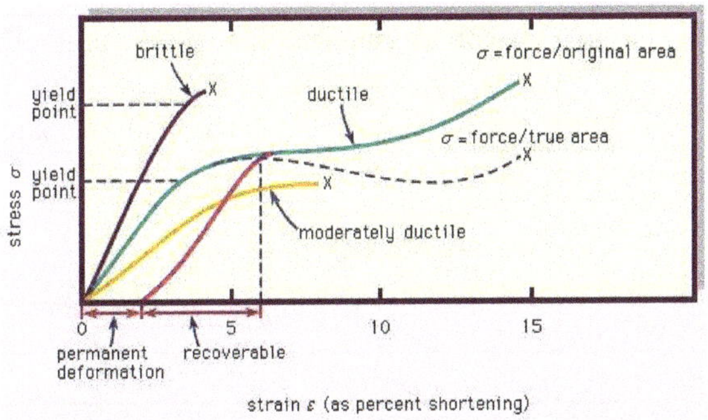

Figure: Typical stress-strain curves for rock materials. Each X represents the point of fracture for the corresponding material.

If the applied stress is removed while a ductile material is in the plastic range, part of the strain is recoverable (elastically), but there is permanent deformation. The ultimate strength is the highest point (stress) on a stress-strain curve, often occurring at fracture (which is the complete loss of cohesion). The strength of a material is its resistance to failure (destruction of structure) by flow or fracture; it is a measure of the stress required to deform a body. Typical compressive strengths (the stress required to cause failure under compression) are given in the table above.

Effect of Environmental Conditions

The behaviour and mechanical properties of rocks depend on a number of environmental conditions. (1) Confining pressure increases the elasticity, strength (e.g., yield point and ultimate fracture stress), and ductility. (2) Internal pore-fluid pressure reduces the effective stress acting on the sample, thus reducing the strength and ductility. The effective, or net, confining pressure is the external hydrostatic pressure minus the internal pore-fluid pressure. (3) Temperature lowers the strength, enhances ductility, and may enhance recrystallization. (4) Fluid solutions can enhance deformation, creep, and recrystallization. (5) Time is an influential factor as well. (6) The rate of loading (i.e., the rate at which stress is applied) influences mechanical properties. (7) Compaction, as would occur with burial to depth, reduces the volume of pore space for sedimentary rocks and the crack porosity for crystalline rocks.

Rocks, which are typically brittle at the Earth's surface, can undergo ductile deformation when buried and subjected to increased confining pressure and temperature for long periods of time. If stress exceeds their strength or if they are not sufficiently ductile, they will fail by fracture—as a crystal, within a bed or rock, on an earthquake fault zone, and so on—whereas with ductility they can flow and fold.

Some strengths for various rock types under different temperatures and confining pressures are listed in the Table. The plastic yield strength here is the stress at a 2 percent strain; the ultimate strength, as stated above, is the highest point (stress) on the stress-strain curve.

Rock strengths, with varying temperature and pressure				
Rock type	Temperature (°C)	Confining pressure (kilobars)	Plastic yield strength (kilobars)	Ultimate strength (kilobars)
Granite	500	5	10	11.5
	800	5	5	6
Gabbro	500	5	4	8
Peridotite	500	5	8	9
	800	5	5.5	8
Basalt	500	5	8	10
	800	5	2	2.5
Marble	24	2	2.5	5.5
	500	3	1	2
Limestone	24	2	4.5	5.5
	500	3	2.5	3
Dolomite	24	2	6	7
	500	5	4	6.5
Shale	24	2	1.5	2.5
Rock salt	24	1	0.5	1

An increase in confining pressure causes brittle fracture to become shear slippage and eventually causes flow (ductile) behaviour. This transition is also aided by higher temperature, decreased internal pore-fluid pressure, and slower strain rate.

The Table gives the values of some elastic constants—bulk modulus (k), Young's modulus (E), shear modulus (μ), and Poisson's ratio (σp)—at room pressure (1 bar) and high confining pressure (3,000 bars). The values for clastic sedimentary rocks would be particularly variable.

Variation of some elastic constants (in 10^6 bars) with rock type and confining pressure				
	at pressure = 1 bar			
Rock type	Bulk modulus	Young's modulus	Shear modulus	Poisson's ratio
Granite	0.1	0.3	0.2	0.05
Gabbro	0.3	0.9	0.6	0.1
Dunite	1.1	1.5	0.5	0.3
Obsidian	0.4	0.7	0.3	0.08
Basalt	0.5	0.8	0.3	0.23
Gneiss	0.1	0.2	0.1	0.05
Marble	0.1	0.4	0.2	0.1
Quartzite				
Sandstone	0.07	0.2	0.08	0.1
Shale	0.04	0.1	0.05	0.04
Limestone	0.8	0.6	0.2	0.3
	at pressure = 3,000 bars			
Rock type	Bulk modulus	Young's modulus	Shear modulus	Poisson's ratio
Granite	0.5	0.6	0.4	0.25
Gabbro	0.9	0.8	0.5	0.2
Dunite	1.2	1.7	0.7	0.27
Obsidian				
Basalt	0.8	1.2	0.4	0.25
Gneiss	0.5	0.7	0.3	
Marble	0.8	0.7	0.3	0.3
Quartzite	0.5	1.0	0.4	0.07
Sandstone				
Shale				
Limestone				

Thermal Properties

Heat flow (or flux), q, in the Earth's crust or in rock as a building material, is the product of the temperature gradient (change in temperature per unit distance) and the material's thermal conductivity (k, the heat flow across a surface per unit area per unit time when a temperature difference exists in unit length perpendicular to the surface). Thus,

$$q = \frac{dt}{dz} \times k,$$

The units of the terms in this equation are given below, expressed first in the centimetre-gram-second (cgs) system and then in the International System of Units (SI) system, with the conversion factor from the first to the second given between them.

q, heat flow :

calories per square centimetre per second (cal/cm². sec)	$\times\frac{1}{23.9}\times 10^{-6} =$	watts per square metre (W/m²)

$\frac{dt}{dz}$, temperature gradient :

Degree Celsius per Kilometre ($^{\circ}C$ /km, Practical unit for Earth)	$\times 10^{-3} =$	degrees Celsius per metre ($^{\circ}C$ /m)

K, thermal conductivity

Calories per square Centimetre per second Per degree Celsius (cal/cm². sec. $^{\circ}C$)	$\times\frac{1}{23.9}\times 10^{-3} =$	watts per metre per degree Celsius (W/m. $^{\circ}C$)

Thermal Conductivity

Thermal conductivity can be determined in the laboratory or in situ, as in a borehole or deep well, by turning on a heating element and measuring the rise in temperature with time. It depends on several factors: (1) chemical composition of the rock (i.e., mineral content), (2) fluid content (type and degree of saturation of the pore space); the presence of water increases the thermal conductivity (i.e., enhances the flow of heat), (3) pressure (a high pressure increases the thermal conductivity by closing cracks which inhibit heat flow), (4) temperature, and (5) isotropy and homogeneity of the rock.

Typical values of thermal conductivities of rock materials are given in the Table. For crystalline silicate rocks—the dominant rocks of the "basement" crustal rocks—the lower values are typical of ones rich in magnesium and iron (e.g., basalt and gabbro) and the higher values are typical of those rich in silica (quartz) and alumina (e.g., granite). These values result because the thermal conductivity of quartz is relatively high, while that for feldspars is low.

Typical values of thermal conductivity (In 0.001 calorie per cm per second per °C)		
Material	at 20 °C	at 200 °C
Typical rocks	4–10	
Granite	7.8	6.6
Gneiss (perpendicular to banding)	5.9	5.5 (100 °C)
Gneiss (parallel to banding)	8.2	7.4 (100 °C)
Gabbro	5.1	5.0
Basalt	4.0	4.0
Dunite	12.0	8.1
Marble	7.3	5.2
Quartzite	15.0	9.0
Limestone	6.0	
One sandstone (dry)	4.4	
One sandstone (saturated)	5.4	
Shale	3–4	
Rock salt	12.8	
Sand (dry)	0.65	
Sand (30% water)	3.94	
Water	1.34 (0 °C)	1.6 (80 °C)
Ice	5.3 (0 °C)	9.6 (–130 °C)
Magnetite	12.6	
Quartz	20.0	
Feldspars	5.0	

Thermal Expansion

The change in dimension—linear or volumetric—of a rock specimen with temperature is expressed in terms of a coefficient of thermal expansion. This is given as the ratio of dimension change (e.g., change in volume) to the original dimension (volume, V) per unit of temperature (T) change:

$$\frac{1}{V}\frac{\Delta V}{\Delta T}.$$

Most rocks have a volume-expansion coefficient in the range of 15–33 × 10^{-6} per degree Celsius under ordinary conditions. Quartz-rich rocks have relatively high values because of the higher volume expansion coefficient of quartz. Thermal-ex-pansion coefficients increase with temperature.

Table lists some linear-expansion coefficients.

Thermal expansion of rocks	
Rock type	linear expansion coefficient ($\times 10^{-6}$ per degree Celsius)
Granite and rhyolite	8 ± 3
Andesite and diorite	7 ± 2
Basalt, gabbro, and diabase	5.4 ± 1
Sandstone	10 ± 2
Limestone	8 ± 4
Marble	7 ± 2
Slate	9 ± 1

$$\frac{1}{L} = \frac{\Delta L}{\Delta T}.$$

Where L represents length. All data are based on at least three samples.

Radioactive Heat Generation

The spontaneous decay (partial disintegration) of the nuclei of radioactive elements provides decay particles and energy. The energy, composed of emission kinetic en-ergy and radiation, is converted to heat; it has been an important factor in affecting the temperature gradient and thermal evolution of the Earth. Deep-seated elevated temperatures provide the heat that causes rock to deform plastically and to move, thus generating to a large extent the processes of plate tectonics—plate motions, seafloor spreading, continental drift, and subduction—and most earthquakes and volcanism.

Some elements, or their isotopes (nuclear species with the same atomic number but different mass numbers), decay with time. These include elements with an atomic number greater than 83—of which the most important are uranium-235, uranium-238, and thorium-232—and a few with a lower atomic number, such as potassium-40.

The heat generated within rocks depends on the types and abundances of the radioactive elements and their host minerals. Such heat production, A, is given in calories per cubic centimetre per second, or 1 calorie per gram per year = 4.186 × 107 ergs per gram per year = 1.327 ergs per gram per second. The rate of radioactive decay, statistically an exponential process, is given by the half-life, t1/2. The half-life is the time required for half the original radioactive atoms to decay for a particular isotope.

Some radioactive decay series are listed in the table below. The isotopic abundance is the percent of the natural element that exists as that particular radioactive isotope; for example, 99.28 percent of natural uranium is U-238, and 100 percent of thorium is the radioactive Th-232. The final product is the end result of the process (usually multistage) of disintegration. The Table gives the heat productivities of radioactive elements and rock types as reported by George D. Garland. For the rocks, the typical content is given for uranium and thorium (in parts per million [ppm] of weight) and for potassium (in weight percent). The heat production of natural uranium is close to that for the isotope U-238, since almost all natural uranium is of that isotopic species.

Heat productivities				
Isotope	Heat productivity, A (calories per gram per year)			
U-235	4.29			
U-238	0.71			
Natural uranium	0.73			
Th-232	0.20			
K-40	0.22			
Natural potassium	$27(10^{-6})$			
Rb-87	$130(10^{-6})$			
Natural rubidium	$36(10^{-6})$			
Major rock province	Concentration			Heat productivity, a
	U (ppm)	Th (ppm)	K (%)	($\times 10^{-13}$ calories per cubic centimetre per second)
Oceanic crust	0.42	1.68	0.69	0.71
Continental shield crust (old)	1.00	4.00	1.63	1.67
Continental upper crust (young)	1.32	5.28	2.15	2.20

Some radioactive decay series				
Element	Radioactive isotope	Final product	Isotopic abundance (%)	Half-life ($\times 10^9$ years)
Uranium	U-235	Pb-207	0.72	0.7
	U-238	Pb-206	99.28	4.5
Thorium	Th-232	Pb-208	100.0	14.0
Potassium	K-40	(89%) Ca-40	0.01	1.4*
		(11%) Argon-40		11.9*
Rubidium	Rb-87	Sr-87	27.8	48.8
*Half-life for K-40 as a whole is 1.25×10^9 years.				

The radioactive elements are more concentrated in the continental upper-crust rocks that are rich in quartz (i.e., felsic, or less mafic). This results because these rocks are differentiated by partial melting of the upper-mantle and oceanic-crust rock. The radioactive elements tend to be preferentially driven off from these rocks for geochemical reasons. A compilation of heat productivities of various rock types is given in the table.

Heat productivities of various rocks								
	Abundances				Heat production			
Rock type	U ppm	Th ppm	Rb ppm	K %	From U	From Th	From K	Total A (in 10^{-6} calories per gram per year)
Granite	3.4	50	220	4.45	2.52	9.95	1.16	13.63
Andesite	1.9	6.4	67	2.35	1.41	1.27	0.61	3.29
Oceanic basalt	0.5	0.9	9	0.43	0.37	0.18	0.11	0.66
Peridotite	0.005	0.01	0.063	0.001	0.0037	0.002	0.0003	0.006
Average upper continental crust	2.5	10.5	110	2.7	1.85	2.09	0.7	4.64
Average continental crust	1.0	2.5	50	1.25	0.74	0.5	0.33	1.56

Electrical Properties

The electrical nature of a material is characterized by its conductivity (or, inversely, its resistivity) and its dielectric constant, and coefficients that indicate the rates of change of these with temperature, frequency at which measurement is made, and so on. For rocks with a range of chemical composition as well as variable physical properties of porosity and fluid content, the values of electrical properties can vary widely.

Resistance (R) is defined as being one ohm when a potential difference (voltage; V) across a specimen of one volt magnitude produces a current (i) of one ampere; that is, $V = Ri$. The electrical resistivity (ρ) is an intrinsic property of the material. In other words, it is inherent and not dependent on sample size or current path. It is related to resistance by $R = \rho L/A$ where L is the length of specimen, A is the cross-sectional area of specimen, and units of ρ are ohm-centimetre; 1 ohm-centimetre equals 0.01 ohm-metre. The conductivity (σ) is equal to $1/\rho$ ohm^{-1}. $centimetre^{-1}$ (or termed mhos/cm). In SI units, it is given in mhos/metre, or siemens/metre.

Some representative values of electrical resistivity for rocks and other materials are listed in the Table. Materials that are generally considered as "good" conductors have a resistivity of 10^{-5}–10 ohm-centimetre (10^{-7}–10^{-1} ohm-metre) and a conductivity of 10–10^{7} mhos/metre. Those that are classified as intermediate conductors have a resistivity of 100–10^{9} ohm-centimetre (1–10^{7} ohm-metre) and a conductivity of 10^{-7}–1 mhos/metre. "Poor" conductors, also known as insulators, have a resistivity of 10^{10}–10^{17} ohm-centimetre (10^{8}–10^{15} ohm-metre) and a conductivity of 10^{-15}–10^{-8}. Seawater is a much better conductor (i.e., it has lower resistivity) than fresh water owing to its higher content of dissolved salts; dry rock is very resistive. In the subsurface, pores are typically filled to some degree by fluids. The resistivity of materials has a wide range—copper is, for example, different from quartz by 22 orders of magnitude.

Typical resistivities		
Material		Resistivity (ohm-centimetre)
Seawater (18 °C)		21
Uncontaminated surface water		$2(10^4)$
Distilled water		$0.2–1(10^6)$
Water (4 °C)		$9(10^6)$
Ice		$3(10^8)$
Rocks in situ		
Sedimentary	clay, soft shale	$100–5(10^3)$
	hard shale	$7–50(10^3)$
	sand	$5–40(10^3)$
	sandstone	$(10^4)–(10^5)$
	glacial moraine	$1–500(10^3)$
	porous limestone	$1–30(10^4)$
	dense limestone	$>(10^6)$
	rock salt	$(10^8)–(10^9)$
Igneous		$5(10^4)–(10^8)$
Metamorphic		$5(10^4)–5(10^9)$
Rocks in laboratory		
Dry granite		10^{12}
Minerals		
Copper (18 °C)		$1.7(10^{-6})$
Graphite		$5–500(10^{-4})$
Pyrrhotite		0.1–0.6
Magnetite crystals		0.6–0.8
Pyrite ore		$1–(10^5)$
Magnetite ore		$(10^2)–5(10^5)$
Chromite ore		$>10^6$
Quartz (18 °C)		$(10^{14})–(10^{16})$

For high-frequency alternating currents, the electrical response of a rock is governed in part by the dielectric constant, ε. This is the capacity of the rock to store electric charge; it is a measure of polarizability in an electric field. In cgs units, the dielectric constant is 1.0 in a vacuum. In SI units, it is given in farads per metre or in terms of the ratio of specific capacity of the material to specific capacity of vacuum (which is 8.85×10^{-12} farads per metre). The dielectric constant is a function of temperature, and of frequency, for those frequencies well above 100 hertz (cycles per second).

Electrical conduction occurs in rocks by (1) fluid conduction—*i.e.*, electrolytic conduction by ionic transfer in briny pore water—and (2) metallic and semiconductor (*e.g.*, some sulfide ores) electron conduction. If the rock has any porosity and contained fluid, the fluid typically dominates the conductivity response. The rock conductivity depends on the conductivity of the fluid (and its chemical composition), degree of fluid saturation, porosity and permeability, and temperature. If rocks lose water, as with compaction of clastic sedimentary rocks at depth, their resistivity typically increases.

Magnetic Properties

The magnetic properties of rocks arise from the magnetic properties of the constituent mineral grains and crystals. Typically, only a small fraction of the rock consists of magnetic minerals. It is this small portion of grains that determines the magnetic properties and magnetization of the rock as a whole, with two results: (1) the magnetic properties of a given rock may vary widely within a given rock body or structure, depending on chemical in homogeneities, depositional or crystallization conditions, and what happens to the rock after formation; and (2) rocks that share the same lithology (type and name) need not necessarily share the same magnetic characteristics. Lithologic classifications are usually based on the abundance of dominant silicate minerals, but the magnetization is determined by the minor fraction of such magnetic mineral grains as iron oxides. The major rock-forming magnetic minerals are iron oxides and sulfides.

Although the magnetic properties of rocks sharing the same classification may vary from rock to rock, general magnetic properties do nonetheless usually depend on rock type and overall composition. The magnetic properties of a particular rock can be quite well understood provided one has specific information about the magnetic properties of crystalline materials and minerals, as well as about how those properties are affected by such factors as temperature, pressure, chemical composition, and the size of the grains. Understanding is further enhanced by information about how the properties of typical rocks are dependent on the geologic environment and how they vary with different conditions.

Applications of the Study of Rock Magnetization

An understanding of rock magnetization is important in at least three different areas: prospecting, geology, and materials science. In magnetic prospecting, one is interested in mapping the depth, size, type, and inferred composition of buried rocks. The prospecting, which may be done from ground surface, ship, or aircraft, provides an important first step in exploring buried geologic structures and may, for example, help identify favourable locations for oil, natural gas, and economic mineral deposits.

Rock magnetization has traditionally played an important role in geology. Paleo magnetic work seeks to determine the remanent magnetization and thereby ascertain the character of the Earth's field when certain rocks were formed. The results of such research have important ramifications in stratigraphic correlation, age dating, and reconstructing past movements of the Earth's crust. Indeed, magnetic surveys of the oceanic crust provided for the first time the quantitative evidence needed to cogently demonstrate that segments of the crust had undergone large-scale lateral displacements over geologic time, thereby corroborating the concepts of continental drift and seafloor spreading, both of which are fundamental to the theory of plate tectonics.

The understanding of magnetization is increasingly important in materials science as well. The design and manufacture of efficient memory cores, magnetic tapes, and permanent magnets increasingly rely on the ability to create materials having desired magnetic properties.

Basic Types of Magnetization

There are six basic types of magnetization:

- Diamagnetism,
- Paramagnetism,
- Ferromagnetism,
- Antiferromagnetism,
- Ferrimagnetism, and
- Superparamagnetism.

Diamagnetism arises from the orbiting electrons surrounding each atomic nucleus. When an external magnetic field is applied, the orbits are shifted in such a way that the atoms set up their own magnetic field in opposition to the applied field. In other words, the induced diamagnetic field opposes the external field. Diamagnetism is present in all materials, is weak, and exists only in the presence of an applied field. The propensity of a substance for being magnetized in an external field is called its susceptibility (k) and it is defined as J/H, where J is the magnetization (intensity) per unit volume and H is the strength of the applied field. Since the induced field always opposes the applied field, the sign of diamagnetic susceptibility is negative. The susceptibility of a diamagnetic substance is on the order of -10^{-6} electromagnetic units per cubic centimetre (emu/cm^3). It is sometimes denoted κ for susceptibility per unit mass of material.

Paramagnetism results from the electron spin of unpaired electrons. An electron has a magnetic dipole moment—which is to say that it behaves like a tiny bar magnet—and so when a group of electrons is placed in a magnetic field, the dipole moments tend to line up with the field. The effect augments the net magnetization in the direction of the applied field. Like diamagnetism, paramagnetism is weak and exists only in the presence of an applied field, but since the effect enhances the applied field, the sign of the paramagnetic susceptibility is always positive. The susceptibility of a paramagnetic substance is on the order of 10^{-4} to 10^{-6}emu/cm^3.

Ferromagnetism also exists because of the magnetic properties of the electron. Unlike paramagnetism, however, ferromagnetism can occur even if no external field is applied. The magnetic dipole moments of the atoms spontaneously line up with one another because it is energetically favourable for them to do so. A remanent magnetization can be retained. Complete alignment of the dipole moments would take place only at a temperature of absolute zero (0 kelvin [K], or -273.15 °C). Above absolute zero, thermal motions begin to disorder the magnetic moments. At a temperature called the Curie temperature, which varies from material to material, the thermally induced disorder overcomes the alignment, and the ferromagnetic properties of the substance disappear. The susceptibility of ferromagnetic materials is large and positive. It is on the order of 10 to 10^4 emu/cm^3. Only a few materials—iron, cobalt, and nickel—are ferromagnetic in the strict sense of the word and have a strong residual magnetization. In general usage, particularly in engineering, the term ferromagnetic is frequently applied to any material that is appreciably magnetic.

Antiferromagnetism occurs when the dipole moments of the atoms in a material assume an antiparallel arrangement in the absence of an applied field. The result is that the sample has no net magnetization. The strength of the susceptibility is comparable to that of paramagnetic materials. Above a temperature called the Néel temperature, thermal motions destroy the antiparallel arrangement, and the material then becomes paramagnetic. Spin-canted (anti)ferromagnetism is a special condition which occurs when antiparallel magnetic moments are deflected from the antiferromagnetic plane, resulting in a weak net magnetism. Hematite (α-Fe_{2O3}) is such a material.

Ferrimagnetism is an antiparallel alignment of atomic dipole moments which does yield an appreciable net magnetization resulting from unequal moments of the magnetic sublattices. Remanent magnetization is detectable. Above the Curie temperature the substance becomes paramagnetic. Magnetite (Fe_3O_4), which is the most magnetic common mineral, is a ferrimagnetic substance.

Superparamagnetism occurs in materials having grains so small (about 100 angstroms) that any cooperative alignment of dipole moments is overcome by thermal energy.

Types of Remanent Magnetization

Rocks and minerals may retain magnetization after the removal of an externally applied field, thereby becoming permanent weak magnets. This property is known as remanent magnetization and is manifested in different forms, depending on the magnetic properties of the rocks and minerals and their geologic origin and history. Delineated below are the kinds of remanent magnetization frequently observed.

CRM (chemical, or crystallization, remanent magnetization) can be induced after a crystal is formed and undergoes one of a number of physicochemical changes, such as oxidation or reduction, a phase change, dehydration, recrystallization, or precipitation of natural cements. The induction, which is particularly important in some (red) sediments and metamorphic rocks, typically takes place at constant temperature in the Earth's magnetic field.

DRM (depositional, or detrital, remanent magnetization) is formed in clastic sediments when fine particles are deposited on the floor of a body of water. Marine sediments, lake sediments, and some clays can acquire DRM. The Earth's magnetic field aligns the grains, yielding a preferred direction of magnetization.

IRM (isothermal remanent magnetization) results from the application of a magnetic field at a constant (isothermal) temperature, often room temperature.

NRM (natural remanent magnetization) is the magnetization detected in a geologic in situ condition. The NRM of a substance may, of course, be a combination of any of the other remanent magnetizations described here.

PRM (pressure remanent, or piezoremanent, magnetization) arises when a material undergoes mechanical deformation while in a magnetic field. The process of deformation may result from hydrostatic pressure, shock impact (as produced by a meteorite striking the Earth's surface), or directed tectonic stress. There are magnetization changes with stress in the elastic range, but the most pronounced effects occur with plastic deformation when the structure of the magnetic minerals is irreversibly changed.

TRM (thermoremanent magnetization) occurs when a substance is cooled, in the presence of a magnetic field, from above its Curie temperature to below that temperature. This form of magnetization is generally the most important, because it is stable and widespread, occurring in igneous and sedimentary rocks. TRM also can occur when dealing exclusively with temperatures below the Curie temperature. In PTRM (partial thermoremanent magnetization) a sample is cooled from a temperature below the Curie point to yet a lower temperature.

VRM (viscous remanent magnetization) results from thermal agitation. It is acquired slowly over time at low temperatures and in the Earth's magnetic field. The effect is weak and unstable but is present in most rocks.

Hysteresis and Magnetic Susceptibility

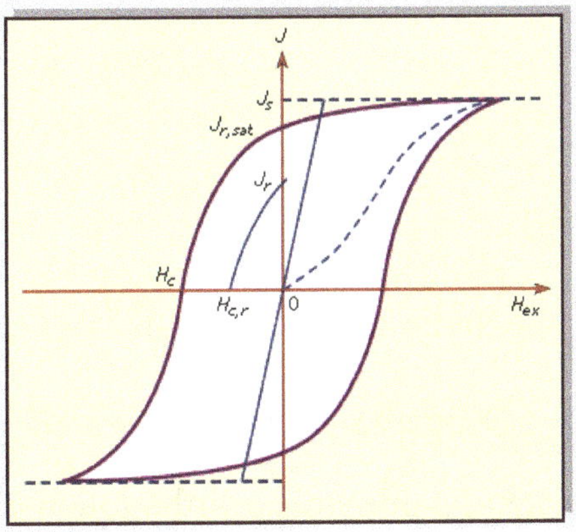

The concept of hysteresis is fundamental when describing and comparing the magnetic properties of rocks. Hysteresis is the variation of magnetization with applied field and illustrates the ability of a material to retain its magnetization, even after an applied field is removed. The figure below illustrates this phenomenon in the form of a plot of magnetization (J) versus applied field (Hex). J_s is the saturation (or "spontaneous")

magnetization when all the magnetic moments are aligned in their configuration of maximum order. It is temperature-dependent, reaching zero at the Curie temperature. $J_{r,sat}$ is the remanent magnetization that remains when a saturating (large) applied field is removed, and Jr is the residual magnetization left by some process apart from IRM saturation, as, for example, TRM. H_c is the coercive field (or force) that is required to reduce $J_{r,sat}$ to zero, and $H_{c,r}$ is the field required to reduce J_r to zero.

General magnetic hysteresis curve, showing magnetization (J) as a function of the external field (Hex). J_s is the saturation (or "spontaneous") magnetization; $J_{r,sat}$ is the remanent magnetization that remains after a saturating applied field is removed; J_r is the residual magnetization left by some magnetization process other than IRM saturation; H_c is the coercive field; and $H_{c,r}$ is the field necessary to reduce J_r to zero.

Magnetic susceptibility is a parameter of considerable diagnostic and interpretational use in the study of rocks. This is true whether an investigation is being conducted in the laboratory or magnetic fields over a terrain are being studied to deduce the structure and lithologic character of buried rock bodies. Susceptibility for a rock type can vary widely, depending on magnetic mineralogy, grain size and shape, and the relative magnitude of remanent magnetization present, in addition to the induced magnetization from the Earth's weak field. The latter is given as $J_{induced} = k_{Hex}$, where k is the (true) magnetic susceptibility and H_{ex} is the external (i.e., the Earth's) magnetic field. If there is an additional remanent magnetization with its ratio (Q_n) to induced magnetization being given by

$$Q_n = \frac{J_{remanent}}{kH_{ex}},$$

then the total magnetization is,

$$\begin{aligned} J &= J_{induced} + J_{remanent} \\ &= kH_{ex} + Q_n kH_{ex} \\ &= k(1+Q_n)H_{ex} \\ &= k_{app}H_{ex}. \end{aligned}$$

Where k_{app}, the "apparent" magnetic susceptibility, is $k(1 + Q_n)$.

Magnetic Minerals and Magnetic Properties of Rocks

The major rock-forming magnetic minerals are the following iron oxides: the titanomagnetite series, $x\text{Fe}_2\text{TiO}_4 \cdot (1 - x)\text{Fe}_{304}$, where Fe_3O_4 is magnetite, the most magnetic mineral; the ilmenohematite series, $y\text{FeTiO}_3 \cdot (1 - y)\text{Fe}_2\text{O}_3$, where $\alpha\text{-Fe}_2\text{O}_3$ (in its rhombohedral structure) is hematite; maghemite, $\gamma\text{-Fe}_2\text{O}_3$ (in which some iron atoms are missing in the hematite structure); and limonite (hydrous iron oxides). They also include sulfides—namely, the pyrrhotite series, $y\text{FeS} \cdot (1 - y)\text{Fe}_{1-x}\text{S}$.

The table below gives some typical values of the apparent susceptibility for various rock types, which usually include some remanent as well as induced magnetization. Values are higher for mafic igneous rocks, especially as the content of magnetite increases.

Approximate "apparent susceptibilities" for rock types	
Rock	Apparent magnetic susceptibility (electromagnetic units per cubic centimetre)
Iron ores	over 0.1
Basalt	10^{-2}
Andesite	10^{-3}
Dacite	10^{-4}
Metamorphic rocks	10^{-4}
Sedimentary rocks	10^{-5}

A distribution of measured (true) susceptibilities for various rock types is shown in the Table. Basic refers to those rocks high in iron and magnesium silicates and magnetite, extrusive means formed by cooling after extruding onto the land surface or seafloor. The data in each category are based on at least 45 samples.

Measured susceptibilities for rock types				
Rock type	% of samples with magnetic susceptibility (in 10^{-6} electromagnetic unit per cubic centimetre)			
	Less than 100	100–1,000	1,000–4,000	Greater than 4,000
Basic extrusive (e.g., basalt)	5	29	47	19
Basic intrusive (e.g., gabbro)	24	27	28	21
Granite	60	23	16	1
Metamorphic (gneiss, schist, slate)	71	22	7	0
Sedimentary	73	19	4	4

The Table lists representative values for the magnetic properties J_n (natural remanent magnetization), k (susceptibility), and ratio Q_n. Natural remanent magnetization is some combination of remanences; typically TRM in an igneous rock, perhaps DRM or CRM or both in a sedimentary rock, and all with an additional VRM. The ratio Q_n is typically higher for rocks with a strong, stable remanence—*e.g.*, magnetite-rich and fine-grained extrusive rocks such as seafloor basalts.

Typical magnetic properties of rocks			
Rock	J_n	k	ratio*
	Natural remanent magnetization (10^{-5} electromagnetic units per cubic centimetre)	Magnetic susceptibility (10^{-5}electromagnetic units per cubic centimetre)	$Q_n=J_n/k \cdot H_{ex}$
Igneous			
Granite	10–80	50–400	0.3–1
Diabase	190–400	100–230	2–3.5

Basalt	200–1,000	100–700	5–10
Seafloor basalt (1–6 metres)	500–800	30–60	25–45
Typical (average)	10–4,000	5–500	1–40
Sedimentary			
Red sediments	0.2–2	0.04–6	2–4
Sandstone		1–40	
Shale		1–50	
Limestone		0.5–20	
Typical (average)	0.1–10	0.3–30	0.02–10
Ores			
Magnetite ore	300,000–1,000,000	30,000–100,000	~10–50
Hematite ore		10–70	
*For external magnetic field (Hex) = 0.5 oersted, the cgs electromagnetic unit of magnetic field intensity.			

Petrology

Petrology is a field of geology that focuses on the study of rocks and the conditions under which they are formed. It utilizes the classical fields of mineralogy, petrography, and chemical analyses to describe the structure and composition of rocks. In addition, modern petrologists include the principles of geochemistry and geophysics to better understand the origins of rocks. There are three branches of petrology, corresponding to the three main types of rocks: igneous, metamorphic, and sedimentary.

A sample of igneous rock. The light-colored tracks show the direction of lava flow.

The study of rocks provides us with important information about the nature of the Earth's crust and mantle. In addition, it enables us to gain a sense of the Earth's history, including tectonic processes that occurred over the long course of geological time. On a practical level, the field of petrology helps us gain an understanding of many of the raw materials we rely on for our sustenance and technological development.

A sample of quartzite, a form of metamorphic rock, from the collection at the Museum of Geology, University of Tartu.

Significance of Studying Rocks

The study of rocks is important for several reasons:

- Their minerals and global chemistry provide information about the composition of the Earth's crust and mantle.
- Their ages can be calculated by various methods of radiometric dating, and a time sequence of geological events can be put together.
- Their features are usually characteristic of a specific tectonic environment, allowing scientists to reconstitute tectonic processes.
- Many rocks host important ores that provide valuable raw materials that we rely on for our sustenance and technological development.

Two types of sedimentary rock, limey shale overlaid by limestone, observed at Cumberland Plateau, Tennessee.

Importance of Petrology

Petrology plays an important role in ascertaining the physical and chemical composition of rocks and the different conditions that influence their formation. Modern petrologists rely on knowledge in mineralogy to help in mapping and sampling of rocks.

Since most rocks constitute various minerals, it becomes easier to study and understand them with background knowledge in mineralogy.

Petrological research is also crucial in helping us understand the nature of the earth's crust and mantle. Over many years, the earth's tectonic processes have shaped the nature of rocks and the topography of the earth's surface. These geological processes are vital in determining the suitability of certain areas for agriculture, industrial or commercial use.

Aside from helping us understand the history of our planet's formation, petrology applies in many areas that directly affect our livelihood. The field of petrology helps us to understand the best raw materials to be used in industries for the manufacture of goods. Additionally, petrology promotes best practices that foster sustenance and technological advancement.

Igneous Petrology

Igneous petrology is a branch that specializes in the scientific study of igneous rocks, their chemical composition and texture. Igneous rocks are rocks formed when magma or molten rock is crystallized to form granite or basalt. Igneous rocks may be formed through crystallization or may involve a different process that cools or solidifies the molten rock. Igneous rocks that are formed beneath the earth's surface form 'intrusive' rocks, also known as plutonic rocks. Rocks found on the earth's surface produce extrusive rocks. About 95 percent of the earth's crust is made up of igneous rocks. However, these rocks do not readily appear on the surface because they are covered by a thin extensive layer of sedimentary and metamorphic rocks.

Metamorphic Petrology

Metamorphism means change in form. In geology the term is used to refer to a solid-state recrystallization of earlier igneous, sedimentary, or metamorphic rocks. There are two main types of metamorphism: (1) contact metamorphism, in which changes induced largely by increase in temperature are localized at the contacts of igneous intrusions; and (2) regional metamorphism, in which increased pressure and temperature have caused recrystallization over extensive regions in mountain belts. Other types of metamorphism include local effects caused by deformation in fault zones, burning oil shales, and thrusted ophiolite complexes; extensive recrystallization caused by high heat flow in mid-ocean ridges; and shock metamorphism induced by high-pressure impacts of meteorites in craters on the Earth and Moon.

Metamorphic petrology is concerned with field relations and local tectonic environments; the description and classification of metamorphic rocks in terms of their texture and chemistry, which provides information on the nature of the premetamorphic material; the study of minerals and their chemistry (the mineral assemblages and their possible reactions), which yields data on the temperatures and pressures at which the rocks recrystallized; and the study of fabrics and the relations of mineral growth to deformation stages and major structures, which provides information about the tectonic conditions under which regional metamorphic rocks formed.

A supplement to metamorphism is metasomatism: the introduction and expulsion of fluids and elements through rocks during recrystallization. When new crust is formed and metamorphosed at a mid-oceanic ridge, seawater penetrates into the crust for a few kilometres and carries much sodium with it. During formation of a contact metamorphic aureole around a granitic intrusion, hydrothermal fluids carrying elements such as iron, boron, and fluorine pass from the granite into the wall rocks. When the continental crust is thickened, its lower part may suffer dehydration and form granulites. The expelled fluids, carrying such heat-producing elements as rubidium, uranium, and thorium migrate upward into the upper crust. Much petrologic research is concerned with determining the amount and composition of fluids that have passed through rocks during these metamorphic processes.

The basic instrument used by the metamorphic petrologist is the petrographic microscope, which allows detailed study and definition of mineral types, assemblages, and reactions. If a heating/freezing stage is attached to the microscope, the temperature of formation and composition of fluid inclusions within minerals can be calculated. These inclusions are remnants of the fluids that passed through the rocks during the final stages of their recrystallization. The electron microprobe is widely used for analyzing the composition of the component minerals. The petrologist can combine the mineral chemistry with data from experimental studies and thermodynamics to calculate the pressures and temperatures at which the rocks recrystallized. By obtaining information on the isotopic age of successive metamorphic events with a mass spectrometer, pressure–temperature–time curves can be worked out. These curves chart the movement of the rocks over time as they were brought to the surface from deep within the continental crust; this technique is important for understanding metamorphic processes. Some continental metamorphic rocks that contain diamonds and coesites (ultrahigh pressure minerals) have been carried down subduction zones to a depth of at least 100 kilometres (60 miles), brought up, and often exposed at the present surface within resistant eclogites of collisional orogenic belts—such as the Swiss Alps, the Himalayas, the Kokchetav metamorphic terrane in Kazakhstan, and the Variscan belt in Germany. These examples demonstrate that metamorphic petrology plays a key role in unraveling tectonic processes in mountain belts that have passed through the plate-tectonic cycle of events.

Sedimentary Petrology

The field of sedimentary petrology is concerned with the description and classification of sedimentary rocks, interpretation of the processes of transportation and deposition of the sedimentary materials forming the rocks, the environment that prevailed at the time the sediments were deposited, and the alteration (compaction, cementation, and chemical and mineralogical modification) of the sediments after deposition.

A banded-iron formation (BIF) rock recovered from the Temagami greenstone belt in Ontario, Canada, and dated to 2.7 billion years ago. Dark layers of iron oxide are intercalated with red chert.

There are two main branches of sedimentary petrology. One branch deals with carbonate rocks, namely limestones and dolomites, composed principally of calcium carbonate (calcite) and calcium magnesium carbonate (dolomite). Much of the complexity in classifying carbonate rocks stems partly from the fact that many limestones and dolomites have been formed, directly or indirectly, through the influence of organisms, including bacteria, lime-secreting algae, various shelled organisms (e.g., mollusks and brachiopods), and by corals. In limestones and dolomites that were deposited under marine conditions, commonly in shallow warm seas, much of the material initially forming the rock consists of skeletons of lime-secreting organisms. In many examples, this skeletal material is preserved as fossils. Some of the major problems of carbonate petrology concern the physical and biological conditions of the environments in which carbonate material has been deposited, including water depth, temperature, degree of illumination by sunlight, motion by waves and currents, and the salinity and other chemical aspects of the water in which deposition occurred.

The other principal branch of sedimentary petrology is concerned with the sediments and sedimentary rocks that are essentially noncalcareous. These include sands and sandstones, clays and claystones, siltstones, conglomerates, glacial till, and varieties of sandstones, siltstones, and conglomerates (e.g., the graywacke-type sandstones and siltstones). These rocks are broadly known as clastic rocks because they consist of distinct particles or clasts. Clastic petrology is concerned with classification, particularly with respect to the mineral composition of fragments or particles, as well as the shapes of particles (angular versus rounded), and the degree of homogeneity of particle sizes. Other main concerns of clastic petrology are the mode of transportation of sedimentary materials, including the transportation of clay, silt, and fine sand by wind; and the transportation of these and coarser materials through suspension in water, through traction by waves and currents in rivers, lakes, and seas, and sediment transport by ice.

Sedimentary petrology also is concerned with the small-scale structural features of sediments and sedimentary rocks. Features that can be conveniently seen in a specimen held in the hand are within the domain of sedimentary petrology. These features include the geometrical attitude of mineral grains with respect to each other, small-scale cross stratification, the shapes and interconnections of pore spaces, and the presence of fractures and veinlets.

Instruments and methods used by sedimentary petrologists include the petrographic microscope for description and classification, X-ray mineralogy for defining fabrics and small-scale structures, physical model flume experiments for studying the effects of flow as an agent of transport and the development of sedimentary structures, and mass spectrometry for calculating stable isotopes and the temperatures of deposition, cementation, and diagenesis. Wet-suit diving permits direct observation of current processes on coral reefs, and manned submersibles enable observation at depth on the ocean floor and in mid-oceanic ridges.

The plate-tectonic theory has given rise to much interest in the relationships between sedimentation and tectonics, particularly in modern plate-tectonic environments—e.g., spreading-related settings (intracontinental rifts, early stages of intercontinental rifting such as the Red Sea, and late stages of intercontinental rifting such as the margins of the present Atlantic Ocean), mid-oceanic settings (ridges and transform faults), subduction-related settings (volcanic arcs, fore-arcs, back-arcs, and trenches), and continental collision-related settings (the Alpine-Himalayan belt and late orogenic basins with molasse [i.e., thick association of clastic sedimentary rocks consisting chiefly of sandstones and shales]). Today many subdisciplines of sedimentary petrology are concerned with the detailed investigation of the various sedimentary processes that occur within these plate-tectonic environments.

References

- Rock-geology, science: britannica.com, Retrieved 2 February, 2019
- Petrology: newworldencyclopedia.org, Retrieved 11 May, 2019
- What-is-petrology: worldatlas.com, Retrieved 9 January, 2019
- Sedimentary-petrology, geology, science: britannica.com, Retrieved 15 March, 2019

2

Igneous Rocks

The rocks which are formed through the cooling and solidification of magma or lava are known as igneous rocks. They occur in a wide range of geological locations such as platforms, basins and extended crust and oceanic crust. A few examples of igneous rocks are obsidian, diorite, andesite and diabase. The topics elaborated in this chapter will help in gaining a better perspective about the different kinds of igneous rocks.

Igneous rock are of various crystalline or glassy rocks formed by the cooling and solidification of molten earth material. Igneous rocks constitute one of the three principal classes of rocks, the others being metamorphic and sedimentary.

Igneous rocks are formed from the solidification of magma, which is a hot (600 to 1,300 °C, or 1,100 to 2,400 °F) molten or partially molten rock material. Earth is composed predominantly of a large mass of igneous rock with a very thin veneer of weathered material—namely, sedimentary rock. Whereas sedimentary rocks are produced by processes operating mainly at Earth's surface by the disintegration of mostly older igneous rocks, igneous—and metamorphic—rocks are formed by internal processes that cannot be directly observed and that necessitate the use of physical-chemical arguments to deduce their origins. Because of the high temperatures within Earth, the principles of chemical equilibrium are applicable to the study of igneous and metamorphic rocks, with the latter being restricted to those rocks formed without the direct involvement of magma.

Magma is thought to be generated within the plastic asthenosphere (the layer of partially molten rock underlying Earth's crust) at a depth below about 60 kilometres (40 miles). Because magma is less dense than the surrounding solid rocks, it rises toward the surface. It may settle within the crust or erupt at the surface from a volcano as a lava flow. Rocks formed from the cooling and solidification of magma deep within the crust are distinct from those erupted at the surface mainly owing to the differences in physical and chemical conditions prevalent in the two environments. Within Earth's deep crust the temperatures and pressures are much higher than at its surface; consequently, the hot magma cools slowly and crystallizes completely, leaving no trace of the liquid magma. The slow cooling promotes the growth of minerals large enough to be identified visually without the aid of a microscope (called phaneritic, from the Greek phaneros, meaning "visible"). On the other hand, magma erupted at the surface is chilled so quickly that the individual minerals have little or no chance to grow. As a result, the rock is either composed of minerals that can be seen only with the aid of a microscope (called aphanitic, from the Greek aphanēs, meaning "invisible") or contains no minerals at all (in the latter case, the rock is composed of glass, which is a highly viscous liquid). This results in two groups: (1) plutonic intrusive igneous rocks that solidified deep within the crust and (2) volcanic, or extrusive, igneous rocks formed at Earth's surface. Some intrusive rocks, known as subvolcanic, were not formed at great depth but were instead injected near the surface where lower temperatures result in a more rapid cooling process; these tend to be aphanitic and are referred to as hypabyssal intrusive rocks.

The deep-seated plutonic rocks can be exposed at the surface for study only after a long period of denudation or by some tectonic forces that push the crust upward or by a combination of the two conditions. (Denudation is the wearing away of the terrestrial surface by processes including weathering and erosion.) Generally, the intrusive rocks have cross-cutting contacts with the country rocks that they have invaded, and in many cases the country rocks show evidence of having been baked and thermally metamorphosed at these contacts. The exposed intrusive rocks are found in a variety of sizes, from small veinlike injections to massive dome-shaped batholiths, which extend for more than 100 square kilometres (40 square miles) and make up the cores of the great mountain ranges.

Extrusive rocks occur in two forms: (1) as lava flows that flood the land surface much like a river and (2) as fragmented pieces of magma of various sizes (pyroclastic materials), which often are blown through the atmosphere and blanket Earth's surface upon settling. The coarser pyroclastic materials accumulate around the erupting volcano, but the finest pyroclasts can be found as thin layers located hundreds of kilometres from the opening. Most lava flows do not travel far from the volcano, but some low-viscosity flows that erupted from long fissures have accumulated in thick (hundreds of metres) sequences, forming the great plateaus of the world (e.g., the Columbia River plateau of Washington and Oregon and the Deccan plateau in India). Both intrusive and extrusive magmas have played a vital role in the spreading of the ocean basin, in the formation of the oceanic crust, and in the formation of the continental margins. Igneous processes have been active since the onset of the formation of Earth some 4.6 billion years ago. Their emanations have provided the water for the oceans, the gases for the primordial oxygen-free atmosphere, and many valuable mineral deposits.

Composition

Chemical Components

The great majority of the igneous rocks are composed of silicate minerals (meaning that the basic building blocks for the magmas that formed them are made of silicon [Si] and oxygen[O]), but minor occurrences of carbonate-rich igneous rocks are found as well. Indeed, in 1960 a sodium carbonate (Na_2CO_3) lava with only 0.05 weight percent silica (SiO_2) was erupted from Ol Doinyo Lengai, a volcano in northern Tanzania, Africa. Because of the limited occurrence of such carbonate-rich igneous rocks, however, the following discussion will consider the chemistry of silicate rocks only.

The major oxides of the rocks generally correlate well with their silica content: those rocks with low silica content are enriched in magnesium oxide (MgO) and iron oxides (FeO, Fe_2O_3, and Fe_3O_4) and are depleted in soda (Na_2O) and potash (K_2O); those with a large amount of silica are depleted in magnesium oxide and iron oxides but are enriched in soda and potash. Both calcium oxide (CaO) and alumina (Al_2O_3) are depleted in the rocks that have a silica content of less than about 45 weight percent, but, above 45 percent, calcium oxide can be as high as 10 percent; this amount decreases gradually as the silica increases. Alumina in rocks that contain more than 45 percent silica is generally above approximately 14 weight percent, with the greatest abundance occurring at an intermediate silica content of about 56 weight percent. Because of the importance of silica content, it has become common practice to use this feature of igneous rocks as a basis for subdividing them into the following groups: silicic or felsic (or acid, an old and discredited but unfortunately entrenched term), rocks having more than 66 percent silica; intermediate, rocks with 55 to 66 percent silica; and subsilicic, rocks containing less than 55 percent silica. The latter may be further divided into two groups: mafic, rocks with 45 to 55 percent silica and ultramafic, those containing less than 45 percent. The subsilicic rocks, enriched as they are in iron (Fe) and magnesium (Mg), are termed femic (from ferrous iron and magnesium), whereas the silicic rocks are referred to as sialic (from silica and aluminum, with which they are enriched) or salic (from silica and aluminum). The terms mafic (from magnesium and ferrous iron) and felsic (feldspar and silica) are used interchangeably with femic and sialic.

The silica content also reflects the mineral composition of the rocks. As the magma cools and begins to crystallize, silica is taken from the magma to be combined with the other cationic oxides to form the silicate minerals. For example, one mole of SiO_2 is combined with one mole of MgO to make the magnesium-rich pyroxene, $MgSiO_3$ (enstatite): $SiO_2 + MgO \rightarrow MgSiO_3$. Two moles of SiO_2 are needed to be combined with one mole each of CaO and Al_2O_3 to make the calcium-rich plagioclase, $CaAl_2Si_2O_8$ (anorthite). However, in a case where magma does not have enough silica relative to the magnesium oxide to produce the pyroxene, the magma will compensate by making a magnesium-olivine (forsterite; Mg_2SiO_4), along with the pyroxene, since the olivine requires only one-half as much silica for every mole of magnesium oxide. On the other hand, a silicic magma may have excess silica such that some will be left after all the silicate minerals were formed from the combination of the oxides; the remaining "free" silica crystallizes as quartz or its polymorphs. The former case usually occurs in subsilicic rocks that characteristically will have silicate minerals like magnesium-olivine, sodium-nepheline ($NaAlSiO_4$, which requires only one mole of silicon for every mole of sodium [Na]), and leucite ($KAlSi_2O_6$, which requires only two moles of silicon to one mole of potassium [K]). These three minerals substitute in part for enstatite, albite ($NaAlSi_3O_8$, requiring three moles of silicon for one mole of sodium), and orthoclase feldspar ($KAlSi_3O_8$, requiring three moles of silicon for one mole of potassium), respectively. Quartz clearly will not be present in these rocks. Minerals such as magnesium-olivine, nepheline, and leucite are termed undersaturated (with respect to silica), and the subsilicic rocks that contain them are termed undersaturated as well. In the case of rocks that have excess silica, the silicic rocks will have quartz and magnesium-pyroxene, which are considered saturated minerals, and the rocks that contain them are termed supersaturated.

Mineralogical Components

The major mineralogical components of igneous rocks can be divided into two groups: felsic (from feldspar and silica) and mafic (from magnesium and ferrous iron). The felsic minerals include quartz, tridymite, cristobalite, feldspars (plagioclase and alkali feldspar), feldspathoids (nepheline and leucite), muscovite, and corundum. Because felsic minerals lack iron and magnesium, they are generally light in colour and consequently are referred to as such or as leucocratic. The mafic minerals include olivine, pyroxenes, amphiboles, and biotites, all of which are dark in colour. Mafic minerals are said to be melanocratic. These terms can be applied to the rocks, depending on the relative proportion of each type of mineral present. In this regard, the term colour index, which refers to the total percentage of the rock occupied by mafic minerals, is useful. Felsic rocks have a colour index of less than 50, while mafic rocks have a colour index above 50. Those rocks that have a colour index above 90 are referred to as ultramafic. These terms are to be used only for the mineralogical content of igneous rocks because they do not necessarily correlate directly with chemical terms. For example, it is common to find a felsic rock composed almost entirely of the mineral plagioclase, but in chemical terms, such a rock is a subsilicic mafic rock. Another example is an igneous rock consisting solely of pyroxene. Mineralogically it would be termed ultramafic, but chemically it is a mafic igneous rock with a silica content of about 50 percent.

The influence of supersaturation and undersaturation on the mineralogy of a rock was noted above. During the crystallization of magmas, supersaturated minerals will not be formed along with undersaturated minerals. Supersaturated minerals include quartz and its polymorphs and a low-calcium orthorhombic pyroxene. These cannot coexist with any of the feldspathoids (e.g.,leucite and nepheline) or magnesium-rich olivine. In volcanic rocks that have been quenched (cooled rapidly) such that only a small part of the magma has been crystallized, it is possible to find a forsterite (magnesium-rich olivine) crystal surrounded by a glass that is saturated or supersaturated. In this case, the outer rim of the olivine may be corroded or replaced by a magnesium-rich pyroxene (called a reaction rim). The olivine was the first to be crystallized, but it was in the process of reacting with the saturated magma to form the saturated mineral when an eruption halted the reaction. Had the magma been allowed to crystallize fully, all the forsterite would have been transformed into the magnesium-rich pyroxene and quartz may have been crystallized.

Accessory minerals present in igneous rocks in minor amounts include monazite, allanite, apatite, garnets, ilmenite, magnetite, titanite, spinel, and zircon. Glass may be a major phase in some volcanic rock but, when present, is usually found in minor amounts. Igneous rocks that were exposed to weathering and circulating groundwater have undergone some degree of alteration. Common alteration products are talc or serpentine formed at the expense of olivine, chlorites replacing pyroxene and amphiboles, iron oxides replacing any mafic mineral, clay minerals and epidote formed from the feldspars, and calcite that may be formed at the expense of any calcium-bearing mineral by interaction with a carbon dioxide (CO_2)-bearing solution. Glass is commonly altered to clay minerals and zeolite. In some cases, however, glass has undergone a devitrification process (in which it is transformed into a crystalline material) initiated by reaction of the glass with water or by subsequent reheating. Common products of devitrification include quartz and its polymorphs, alkali feldspar, plagioclase, pyroxene, zeolite, clays, and chlorite.

Textural Features

The texture of an igneous rock normally is defined by the size and form of its constituent mineral grains and by the spatial relationships of individual grains with one another and with any glass that may be present. Texture can be described independently of the entire rock mass, and its geometric characteristics provide valuable insights into the conditions under which the rock was formed.

Crystallinity

Among the most fundamental properties of igneous rocks are crystallinity and granularity, two terms that closely reflect differences in magma composition and the differences between volcanic and various plutonic environments of formation. Crystallinity generally is described in terms of the four categories shown in the table below.

Crystallinity categories of igneous rocks	
Crystallinity	Rock term
Entirely crystalline	Holocrystalline
Crystalline material and subordinate glass	Hemicrystalline or hypocrystalline
Glass and subordinate crystalline material	Hemihyaline or hypohyaline
Entirely glassy	Holohyaline or hyaline

Those holocrystalline rocks in which mineral grains can be recognized with the unaided eye are called phanerites, and their texture is called phaneritic. Those with mineral grains so small that their outlines cannot be resolved without the aid of a hand lens or microscope are termed aphanites, and their texture is termed aphanitic. Aphanitic rocks are further described as either microcrystalline or cryptocrystalline, according to whether or not their individual constituents can be resolved under the microscope. The subaphanitic, or hyaline, rocks are referred to as glassy, or vitric, in terms of granularity.

Aphanitic and glassy textures represent relatively rapid cooling of magma and, hence, are found mainly among the volcanic rocks. Slower cooling, either beneath Earth's surface or within very thick masses of lava, promotes the formation of crystals and, under favourable circumstances of magma composition and other factors, their growth to relatively large sizes. The resulting phaneritic rocks are so widespread and so varied that it is convenient to specify their grain size as shown in the table below.

Categories of rock grain size		
Terms in common use	General grain size	
	Igneous rocks in general	pegmatites
Fine-grained	<1 mm	<1 inch
Medium-grained	1–5 mm	1–4 inches
Coarse-grained	5 mm–2 cm	4–12 inches
Very coarse-grained	>2 cm	>12 inches

Granularity

Grain Size

The general grain size ordinarily is taken as the average diameter of dominant grains in the rock; for the pegmatites, which are special rocks with extremely large crystals, it can refer to the maximum exposed dimensions of dominant grains. Most aphanitic rocks are characterized by mineral grains less than 0.3 millimetre (0.01 inch) in diameter, and those in which the average grain size is less than 0.1 millimetre (0.004 inch) are commonly described as dense.

Fabric

A major part of rock texture is fabric or pattern, which is a function of the form and outline of its constituent grains, their relative sizes, and their mutual relationships in space. Many specific terms have been employed to shorten the description of rock fabrics, and even the sampling offered here may seem alarmingly extensive. It should be noted, however, that fabric provides some of the most useful clues to the nature and sequence of magmatic crystallization.

The degree to which mineral grains show external crystal faces can be described as euhedral or panidiomorphic (fully crystal-faced), subhedral or hypidiomorphic (partly faced), or anhedral or allotriomorphic (no external crystal faces). Quite apart from the presence or absence of crystal faces, the shape, or habit, of individual mineral grains is described by such terms as equant, tabular, platy, elongate, fibrous, rodlike, lathlike, needlelike, and irregular. A more general contrast can be drawn between grains of equal (equant) and inequal dimensions. Even-grained, or equigranular, rocks are characterized by essential minerals that all exhibit the same order of grain size, but this implied equality need not be taken too literally. For such rocks the combination terms panidiomorphic-granular, hypidiomorphic-granular, and allotriomorphic-granular are applied according to the occurrence of euhedral, subhedral, and anhedral mineral grains within them. Many fine-grained allotriomorphic-granular rocks are more simply termed sugary, saccharoidal, or aplitic.

Rocks that are unevenly grained, or inequigranular, are generally characterized either by a seriate fabric, in which the variation in grain size is gradual and essentially continuous, or by a porphyritic fabric, involving more than one distinct range of grain sizes. Both of these kinds of texture are common. The relatively large crystals in a porphyritic rock ordinarily occur as separate entities, known as phenocrysts, set in a groundmass or matrix of much finer-grained crystalline material or glass. Quite commonly in many volcanic rocks, phenocrysts are aggregated. When this is observed, the term glomeroporphyritic is used to describe the texture, and the aggregate is referred to as a glomerocryst. In some cases, such glomerocrysts are monomineralic, but more commonly they are composed of two or more minerals. Based on chemical composition, texture, and other criteria such as isotopic analysis, it has been demonstrated that some phenocrysts and glomerocrysts were not crystallized from the host magma but rather were accidentally torn from the country rock by the magma as it rose to the surface. When this has occurred, these phenocrysts are referred to as xenocrysts, while the aggregates can be termed xenoliths. The size of phenocrysts is essentially independent of their abundance relative to the groundmass, and they range in external form from euhedral to anhedral. Most of them are best described as subhedral. Because the groundmass constituents span almost the full ranges of crystallinity and granularity, porphyritic fabric is abundantly represented among the phaneritic, aphanitic, and glassy rocks.

The sharp break in grain size between phenocrysts and groundmass reflects a corresponding change in the conditions that affected the crystallizing magma. Thus, the phenocrysts of many rocks probably grew slowly at depth, following which the nourishing magma rose to Earth's surface as lava, cooled much more rapidly, and congealed to form a finer-grained or glassy groundmass. A porphyritic volcanic rock with a glassy groundmass is described as having a vitrophyric texture and the rock can be called a vitrophyre. Other porphyritic rocks may well reflect less drastic shifts in position and perhaps more subtle and complex changes in conditions of temperature, pressure, or crystallization rates. Many phenocrysts could have developed at the points where they now occur, and some may represent systems with two fluid phases, magma and coexisting gas. Appraisals of the composition of phenocrysts, their distribution, and their periods of growth relative to the accompanying groundmass constituents are important to an understanding of many igneous processes.

Important Textural Types

The articulation of mineral grains is described in terms of planar, smoothly curved, sinuous, sutured, interlocked, or irregular surfaces of mutual boundary. The distribution and orientation of mineral grains and of mineral grains and glass are other elements of fabric that can be useful in estimating the conditions and sequence of mineral formation in igneous rocks.

The following are only a few of the most important examples:

- Directive textures are produced by the preferred orientation of platy, tabular, or elongate mineral grains to yield grossly planar or linear arrangements; they are generally a result of magmatic flowage.
- Graphic texture refers to the regular intergrowth of two minerals, one of them generally serving as a host and the other appearing on surfaces of the host as striplike or cuneiform units with grossly consistent orientation; the graphic intergrowth of quartz in alkali feldspar is a good example.
- Ophitic texture is the association of lath-shaped euhedral crystals of plagioclase, grouped radially or in an irregular mesh, with surrounding or interstitial large anhedral crystals of pyroxene; it is characteristic of the common rock type known as diabase.
- Poikilitic texture describes the occurrence of one mineral that is irregularly scattered as diversely oriented crystals within much larger host crystals of another mineral.
- Reaction textures occur at the corroded margins of crystals, from the corrosive rimming of crystals of one mineral by finer-grained aggregates of another, or as a result of other features that indicate partial removal of crystalline material by reaction with magma or other fluid.

Pyroclastic texture results from the explosive fragmentation of volcanic material, including magma (commonly the light, frothy pumice variety and glass fragments called shards), country rock, and phenocrysts. Fragments less than 2 millimetres in size are called ash, and the rock formed of these is called tuff; fragments between 2 and 64 millimetres are lapilli and the rock is lapillistone; fragments greater than 64 millimetres are called bombs if rounded or blocks if angular, and the corresponding rock is termed agglomerate or pyroclastic breccia, respectively. Commonly, many of these pyroclastic rocks have been formed by dense hot clouds that hug the ground and behave much like a lava flow and hence are given the name pyroclastic flow. Most of these flows are composed of ash-size material; therefore, they are called ash flows and the rocks deposited by them are called ash-flow tuffs. A more general term for rocks deposited by these flows that does not specify size of fragments is ignimbrite. Ash-flow tuffs and other ignimbrites often have zones in which the fragments have been welded. These zones are termed welded tuffs and display a directive planar texture (called eutaxitic) that results from compaction and flattening of pumice fragments. Such pyroclastic flows were responsible for many of the deposits of the eruption of Mount St. Helens in Washington State, U.S. Most eruptions eject fragments that are borne by the wind and deposited subaerially (on the land surface). These deposits are said to be ash-fall tuffs and are recognized by their lamination (formation in thin layers that differ in grain size or composition). They commonly blanket the topography in contrast to the ash-flow deposits, which flow around topographic highs and which are completely unsorted.

Replacement textures occur where a mineral or mineral aggregate has the external crystal form of a preexisting different mineral (pseudomorphism) or where the juxtaposition of two minerals indicates that one was formed at the expense of the other.

Finally, crystal zoning describes faintly to very well-defined geometric arrangements of portions within individual crystals that differ significantly in composition (or some other property) from adjacent portions; most common are successive shells grouped concentrically about the centres of crystals, presumably reflecting shifts in conditions during crystal growth.

Structural Features

The structure of an igneous rock is normally taken to comprise the mutual relationships of mineral or mineral-glass aggregates that have contrasting textures, along with layering, fractures, and other larger-scale features that transect or bound such aggregates. Structure often can be described only in relation to masses of rock larger than a hand specimen, and most of its individual expressions can be closely correlated with physical conditions that existed when the rock was formed.

Small-scale Structural Features

Among the most widespread structural features of volcanic rocks are the porelike openings left by the escape of gas from the congealing lava. Such openings are called vesicles, and the rocks in which they occur are said to be vesicular. Where the openings lie close together and form a large part of the containing rock, they impart to it a slaglike, or scoriaceous, structure. Their relative abundance is even greater in the type of sialic glassy rock known as pumice, which is essentially a congealed volcanic froth. Most vesicles can be likened to peas or nuts in their ranges of size and shape; those that were formed when the lava was still moving tend to be flattened and drawn out in the direction of flow. Others are cylindrical, pearlike, or more irregular in shape, depending in part on the manner of escape of the gas from the cooling lava; most of the elongate ones occur in subparallel arrangements.

Many vesicles have been partly or completely filled with quartz, chalcedony, opal, calcite, epidote, zeolites, or other minerals. These fillings are known as amygdules, and the rock in which they are present is amygdaloidal. Some are concentrically layered, others also include centrally disposed series of horizontal layers, and still others are featured by central cavities into which well-formed crystals project.

Spherulites are light-coloured subspherical masses that commonly consist of tiny fibres and plates of alkali feldspar radiating outward from a centre. Most range from pinpoint to nut size, but some are as much as several feet in diameter. The relatively large ones tend to be internally complex and to contain concentric shells of feldspar fibres with or without accompanying quartz, tridymite, or glass. Spherulites occur mainly in glassy volcanic rocks; they also are present in some partly or wholly crystalline rocks that include shallow-seated intrusive types. Many evidently are products of rapid crystallization, perhaps at points of gas concentration in the freezing magmas. Others, in contrast, were formed more slowly, by devitrification of volcanic glasses, presumably not long after they congealed and while they were still relatively hot.

Lithophysae, also known as stone bubbles, consist of concentric shells of finely crystalline alkali feldspar separated by empty spaces; thus, they resemble an onion or a newly blooming rose. Commonly associated with spherulites in glassy and partly crystalline volcanic rocks of salic composition, many lithophysae are about the size of walnuts. They have been ascribed to short episodes of rapid crystallization, alternating with periods of gas escape when the open spaces were developed by thrusting the feldspathic shells apart or by contraction associated with cooling. The curving cavities commonly are lined with tiny crystals of quartz, tridymite, feldspar, topaz, or other minerals deposited from the gases.

Some glassy rocks of silicic composition are marked by domains of strongly curved, concentrically disposed fractures that promote breakage into rounded masses of pinhead to walnut size. Because their surfaces often have a pearly or shiny lustre, the name perlite is applied to such rocks. Perlite is most common in glassy silicic rocks that have interacted with water to become hydrated. During the hydration process, water enters the glass, breaking the silicon-oxygen bonds and causing an expansion of the glass structure to form the curved cracks. The extent of hydration of glass, indicating the amount of perlite that has been formed from the glass, depends on the climate and on time. In a given area where the climate is expected to be consistent, the thickness of the hydration of the glass surface has been used by archaeologists to date artifacts such as arrowheads composed of the dark volcanic glass known as obsidian and made by early native Americans.

Numerous structural features of comparably small scale occur among the intrusive rocks; these include miarolitic, orbicular, plumose, and radial structures. Miarolitic rocks are felsic phanerites distinguished by scattered pods or layers, ordinarily several centimetres in maximum thickness, within which their essential minerals are coarser-grained, subhedral to euhedral, and otherwise pegmatitic in texture. Many of these small interior bodies, called miaroles, contain centrally disposed crystal-lined cavities that are known as druses or miarolitic cavities. An internal zonal disposition of minerals also is common, and the most characteristic sequence is alkali feldspar with graphically intergrown quartz, alkali feldspar, and a central filling of quartz. Miarolitic structure probably represents local concentration of gases during very late stages in consolidation of the host rocks.

The term orbicular is applied to rounded, onionlike masses with distinct concentric layering that are distributed in various ways through otherwise normal-appearing phaneritic rocks of silicic to mafic composition. The layers within individual masses are typically thin, irregular, and sharply defined, and each differs from its immediate neighbours in composition or texture. Some layers contain tabular or prismatic mineral grains that are oriented radially with respect to the containing orbicule and, hence, are analogous to spherulitic layers in volcanic rocks. The minerals of most orbicules are the same as those of the enclosing rock, but they are not necessarily present in the same proportions. The concentric structure appears to reflect rhythmic crystallization about specific centres, commonly at early stages in consolidation of the general rock mass.

The normal fabric of some relatively coarse-grained plutonic rocks is interrupted by clusters of crystals with radial grouping but without concentric layering. A characteristic plumelike, spraylike, or rosettelike structure is imparted by the markedly elongate form of the participating crystals or crystal aggregates, which seem to have developed outward from common centres by direct crystallization from magma or by replacement of preexisting solid material.

Large-scale Structural Features

Many kinds of larger-scale features occur among both the intrusive and the extrusive rocks. Most of these are mentioned later in connection with rock occurrence or are discussed in other articles, but several are properly introduced here:

Clastic Structures

These are various features that express the accumulation of fragments or the rupturing and dislocation of solid material. In volcanic environments they generally result from explosive activity or the incorporation of solid fragments by moving lava; as such, they characterize the pyroclastic rocks. Among the plutonic rocks, they appear chiefly as local to very extensive zones of pervasive shearing, dislocation, and granulation, commonly best recognized under the microscope. Those developed prior to final consolidation of the rock are termed protoclastic; those developed after final consolidation, cataclastic.

Flow Structures

These are planar or linear features that result from flowage of magma with or without contained crystals. Various forms of faintly to sharply defined layering and lining typically reflect compositional or textural inhomogeneities, and they often are accentuated by concentrations or preferred orientation of crystals, inclusions, vesicles, spherulites, and other features.

Fractures

These are straight or curving surfaces of rupture directly associated with the formation of a rock or later superimposed upon it. Primary fractures generally can be related to emplacement or to subsequent cooling of the host rock mass. The columnar jointing found in many mafic volcanic rocks is a typical result of contraction upon cooling.

Inclusions

These are rounded to angular masses of solid material enclosed within a rock of recognizably different composition or texture. Those consisting of older material not directly related to that of their host are known as xenoliths, and those representing broken-up and detached older parts of the same igneous body that encloses them are termed cognate xenoliths or autoliths.

Pillow Structures

These are aggregates of ovoid masses, resembling pillows or grain-filled sacks in size and shape, that occur in many basic volcanic rocks. The masses are separated or interconnected, and each has a thick vesicular crust or a thinner and more dense glassy rind. The interiors ordinarily are coarser-grained and less vesicular. Pillow structure is formed by rapid chilling of highly fluid lava in contact with water or water-saturated sediments, accompanied by the development of budlike projections with tough, elastic crusts. As additional lava is fed into each bud, it grows into a pillow and continues to enlarge until rupture of the skin permits escape of fresh lava to form a new bud and a new pillow.

Segregations

These are special types of inclusions that are intimately related to their host rocks and in general are relatively rich in one or more of the host-rock minerals. They range from small pods to extensive layers and from early-stage crystal accumulations formed by gravitational settling in magma to very late-stage concentrations of coarse-grained material developed in place.

Zonal Structures

These are arrangements of rock units with contrasting composition, or texture, in an igneous body, commonly in a broadly concentric pattern. Chilled margins, the fine-grained or glassy edges along the borders of many extrusive and shallow-seated intrusive bodies, represent quenching of magma along contacts with cooler country rock. Other kinds of zones generally reflect fractional crystallization of magma and are useful in tracing courses of magmatic differentiation, as will be noted later.

An interesting type of zonal structure is an orbicular configuration that has alternating light and dark repeating bands in an oval arrangement found in some diorites and granodiorites. Pegmatites also often have zonal structures due to fluctuations in fluid composition. This results in "pockets" that may contain gems or other unusual minerals.

Classification of Igneous Rocks

Igneous rocks are classified on the basis of mineralogy, chemistry, and texture. As discussed earlier, texture is used to subdivide igneous rocks into two major groups: (1) the plutonic rocks, with mineral grain sizes that are visible to the naked eye, and (2) the volcanic and hypabyssal types, which are usually too fine-grained or glassy for their mineral composition to be observed without the use of a petrographic microscope. Being rather coarsely grained, phaneritic rocks readily lend themselves to a classification based on mineralogy since their individual mineral components can be discerned, but the volcanic rocks are more difficult to classify because either their mineral composition is not visible or the rock has not fully crystallized owing to fast cooling. As a consequence, various methods employ chemical composition as the criterion for volcanic igneous rock classification. A commonly used technique was introduced at the beginning of the 20th century by the American geologists C. Whitman Cross, Joseph P. Iddings, Louis V. Pirsson, and Henry S. Washington. In this method, the mineral composition of the rock is recalculated into a standard set of typically occurring minerals that theoretically could have developed from the complete equilibrium crystallization at low temperatures of a magma of the indicated bulk composition. The calculated hypothetical mineral composition is called the norm, and the minerals constituting the standard set are termed normative minerals, since they are ordinarily found in igneous rocks. The rock under analysis may then be classified according to the calculated proportions of the normative minerals.

Because other methods for calculating the norm have been devised, this original norm is referred to as the CIPW norm after the initials of the four petrologists who devised the system. The norm calculation allows the petrologist studying an aphanitic rock to "see" the mineral assemblage that corresponds well with the actual mineral assemblage of a plutonic rock of the same composition that had crystallized under equilibrium conditions. Moreover, the norm has been shown to have a thermodynamic basis. The concept of silica saturation discussed above is incorporated into the norm, which will show whether a magma of a certain composition is supersaturated, saturated, or undersaturated by the presence or absence of normative minerals such as quartz, orthopyroxene, olivine, and the feldspathoids.

Classification of Plutonic Rocks

A plutonic rock may be classified mineralogically based on the actual proportion of the various minerals of which it is composed (called the mode). In any classification scheme, boundaries between classes are set arbitrarily; however, if the boundaries can be placed closest to natural divisions or gaps between classes, they will seem less random and subjective, and the standards will facilitate universal understanding. In order to set boundaries nearest to the population lows (of constituent minerals) and to achieve an international consensus, a poll among the world's petrologists was conducted and a modal classification for plutonic igneous rocks was devised. Based mainly on this poll, the International Union of Geological Sciences (IUGS) Subcommission on the Systematics of Igneous Rocks in 1973 suggested the use of the modal composition for all plutonic igneous rocks with a colour index less than 90 and for those plutonic ultramafic rocks with a colour index greater than 90.

The plotting of rock modes on these triangular diagrams is simpler than it may appear. If the colour index is less than 90 and quartz (Q) is present, then the three components, Q + A (alkali feldspar) + P (plagioclase), are recalculated from the mode to sum to 100 percent. Each component is represented by the corners of the equilateral triangle, the length of whose sides are divided into 100 equal parts. Any composition plotting at a corner, therefore, has a mode of 100 percent of the corresponding component. Any point on the sides of the triangle represents a mode composed of the two adjacent corner components. For example, a rock with 60 percent Q and 40 percent A will plot on the QA side at a location 60 percent of the distance from A to Q. A rock containing all three components will plot within the triangle. Since the sides of the triangle are divided into 100 parts, a rock having a mode of 20 percent Q and 80 percent A + P (in unknown proportions for the moment) will plot on the line that parallels the AP side and lies 20 percent of the distance toward Q from the side AP. If this same rock has 30 percent P and 50 percent A, the rock mode will plot at the intersection of the 20 percent Q line described above, with a line paralleling the QA side at a distance 30 percent toward P from the QA side. The third intersecting line for the point is necessarily the line paralleling the QP side at 50 percent of the distance from the side QP toward A.

A rock with 25 percent Q, 35 percent P, and 40 percent A plots in the granite field, whereas one with 25 percent Q, 60 percent P, and 15 percent A plots in the granodiorite field. The latter is close to the average composition of the continental crust of Earth. Igneous rocks normally do not exceed about 50 percent quartz, and the feldspathoidal rocks are relatively rare. The most common plutonic rocks are those in fields numbered 3, 4, 5, 8, 9, 10, and 15. These are found in what have been called granite (used in a loose sense) batholiths, which are irregularly shaped large bodies covering an area greater than 100 square kilometres. Batholiths constitute the cores of the great mountain ranges, such as the Rockies in western North America and the Sierra Nevada in California, U.S. Typically these batholiths are composites of smaller intrusions, each of which may display several different rock types. The average composition is close to that of a granodiorite, but in many batholiths the sequence of intrusions progresses from basic to acidic, with gabbro or quartz diorite being emplaced first. In the Sierra Nevada batholith, the dominant rocks are quartz monzonite and granodiorite, with intrusions including quartz diorite in the far western rim and granite in the east. Batholiths contain medium- to coarse-grained rocks with hypidiomorphic-granular texture. The rocks are generally leucocratic; diorites and quartz diorites typically contain less than 30 percent mafic minerals—e.g., hornblende and biotite. Pyroxenes are rare but are more commonly found in the gabbros. Mineralogically the ratio of hornblende to biotite, the colour index, the calcium content of the plagioclase feldspar, and the ratio of plagioclase to alkali feldspar decrease from diorite to quartz diorite to granodiorite and granite. Common accessory minerals include apatite, titanite, and an opaque mineral such as magnetite or ilmenite.

Ideally it would be preferable to use the same modal scheme for volcanic rocks. This is recommended whenever possible; hence, for this purpose, the volcanic or hypabyssal equivalent of the plutonic rocks are listed in parentheses. It should be noted that the dividing lines and boxes are identical to those for the plutonic rocks.

Classification of Volcanic and Hypabyssal Rocks

Owing to the aphanitic texture of volcanic and hypabyssal rocks, their modes cannot be readily determined; consequently, a chemical classification is widely accepted and employed by most petrologists. One popular scheme is based on the use of both chemical components and normative mineralogy. Because most lay people have little access to analytic facilities that yield igneous rock compositions, only an outline will be presented here in order to provide an appreciation for the classification scheme.

The first major division is based on the alkali (soda + potash) and silica contents, which yield two groups, the subalkaline and alkaline rocks. The subalkaline rocks have two divisions based mainly on the iron content, with the iron-rich group called the tholeiitic series and the iron-poor group called calc-alkalic. The former group is most commonly found along the oceanic ridges and on the ocean floor; the latter group is characteristic of the volcanic regions of the continental margins. In some magmatic arcs (groups of islands arranged in a curved pattern), notably Japan, both the tholeiitic and calc-alkalic series occur. This is the case, for example, in the volcanoes of northeastern Honshu, the largest of Japan's four main islands, and both series may be found within the same volcano. The alkaline rocks frequently occur on oceanic islands (usually formed during the late stages of magma consolidation after tholeiitic eruptions) and in continental rifts (extensive fractures). Based on the relative proportions of soda and potash, the calc-alkalic series is subdivided into the sodic and potassic series.

Chemically the subalkaline rocks are saturated with respect to silica; consequently, they have normative minerals such as orthopyroxene [$Mg(Fe)_2Si_2O_6$] and quartz but lack nepheline and olivine (in the presence of quartz). This chemical property also is reflected in the mode of the basic members that have two pyroxenes, orthopyroxene and augite [$Ca(Mg, Fe)Si_2O_6$], and perhaps quartz. Plagioclase is common in phenocrysts, but it can also occur in holocrystalline rocks in the microcrystalline matrix along with the pyroxenes and an iron–titanium oxide phase. In addition to the differences in iron content between the tholeiitic and calc-alkalic series, the latter has a higher alumina content (16 to 20 percent), and the range in silica content is larger (48 to 75 percent compared to 45 to 63 percent for the former). Hornblende and biotite phenocrysts are common in the calc-alkalic andesites and dacites but are lacking in the tholeiites except as alteration products. The dacites and rhyolites commonly have phenocrysts of plagioclase, alkali feldspar (usually sanidine), and quartz in a glassy matrix. Hornblende and plagioclase phenocrysts are more widespread in dacites than in rhyolites, which have more biotite and alkali feldspar. When occurring near volcanic vents, (openings from which volcanic materials are brought to Earth's surface), basalts and andesites of both series are found as tuffs or agglomerates; otherwise, they typically occur as flows. Dacite and rhyolite occur as flows near vents but are most commonly found as tuffs composed of fragmented pieces of glass, phenocrysts, and rock.

The alkaline rocks typically are chemically undersaturated with respect to silica; hence they lack normative orthopyroxene (i.e., they have only one pyroxene, the calcium-rich augite) and quartz but have normative nepheline. Microscopic examination of the alkali olivine basalts usually reveals phenocrysts with an abundance of olivine, one pyroxene (augite, which is usually titanium-rich), and plagioclase. Nepheline may be seen in the matrix. Trachytes typically are leucocratic with an abundance of feldspars aligned roughly parallel to the direction of the lava flow.

Forms of Occurrence

Extrusive Igneous Rocks

Extrusive igneous rocks are the products of volcanic activity. They appear at the surface as molten lava that spreads in sheets and hardens, or they are made up of fragments of magma ejected from vents by violent gaseous explosions. Large-scale extrusive features include stratovolcanoes (composite cones), shield volcanoes, lava domes, and cinder cones. Smaller extrusive features include lava flows known as pahoehoe and aa.

Intrusive Igneous Rocks

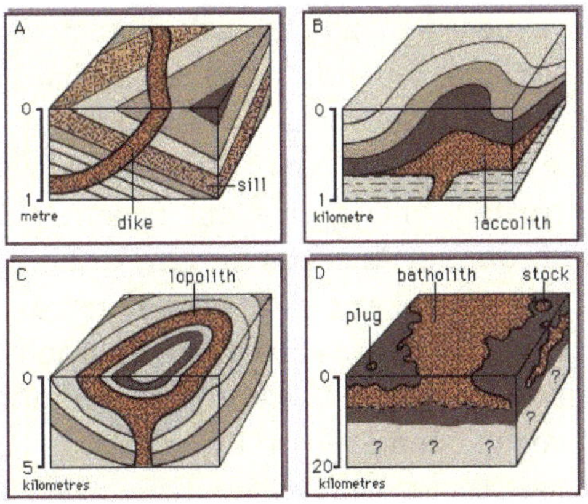

Figure: Forms of intrusive igneous rock bodies in hypothetical sections of Earth strata.

Erosion of volcanoes will immediately expose shallow intrusive bodies such as volcanic necks and diatremes. A volcanic neck is the "throat" of a volcano and consists of a pipelike conduit filled with hypabyssal rocks. Ship Rock in New Mexico and Devil's Tower in Wyoming are remnants of volcanic necks, which were exposed after the surrounding sedimentary rocks were eroded away. Many craterlike depressions may be filled with angular fragments of country rock (breccia) and juvenile pyroclastic debris. When eroded, such a depression exposes a vertical funnel-shaped pipe that resembles a volcanic neck with the exception of the brecciated filling. These pipes are dubbed diatremes. Many diatremes are formed by explosion resulting from the rapid expansion of gas—carbon dioxide and water vapour. These gases are released by the rising magma owing to the decrease in pressure as it nears the surface. Some diatremes contain kimberlite, a peridotite that contains a hydrous mineral called phlogopite. Kimberlite may contain diamonds.

Dikes are usually tabular bodies that may radiate from the central vent of a volcano or from a volcanic neck. Not all dikes are associated with volcanoes, but they can be distinguished by their discordant relationship with the structure of the country rock that they cut across. Many dikes are only a few metres wide, but large ones, such as the dike that feeds the Muskox intrusion in the Northwest Territories of Canada, reach widths of more than 150 metres. Related to dikes are features that maintain a concordant relationship with the structure of the country rocks. Magmas may force their way between layers of country rock and solidify parallel to them to form sills. On the west bank of the Hudson River opposite New York City, the 300-metre-thick Palisades sill is exposed and can be traced for 80 kilometres. A laccolith also is concordant with country rock, but it is distinguished from a sill by having a flat floor with a domed (mushroom-shaped) roof. Laccoliths were first described in the Henry Mountains of Utah, where they may measure up to 200 metres thick with basal diameters exceeding three kilometres. Rocks of intermediate silica content generally make up these domed intrusions. In contrast, lopoliths are saucer-shaped bodies with a concave upward roof and floor and are commonly composed of mafic rocks. Lopoliths are huge in size; the Bushveld intrusive complex in South Africa, for example, has an area of about 66,000 square kilometres and an exposed thickness of 8 kilometres. The Muskox intrusion, mentioned above, is another large lopolith, which is estimated to be about 80 kilometres long and 11 kilometres wide (roof rocks covering part of the intrusion prevent an exact measurement). These lopoliths are commonly layered with igneous minerals and rocks; in the Bushveld intrusion, one layer about 1 metre thick consisting of almost pure chromite (an ore of chromium) extends for tens of kilometres. Large irregularly shaped plutons are called either stocks or batholiths, depending on their sizes. Plutons larger than 100 square kilometres in area are termed batholiths, while those of lesser size are called stocks. It may be possible, however, that some stocks are the visible portions of batholiths that have not been exposed by erosion.

Batholiths are deep-seated crustal intrusions, whereas stocks may be formed at shallow depths only a few kilometres below the surface. Rocks ranging from quartz diorite to granite are commonly found in batholiths. Large batholiths in North America include the Sierra Nevada, the Idaho, and the Coast Range, which is about 600 kilometres long and 200 kilometres wide and extends from the Alaskan border through British Columbia to Washington state. Many pulses of intrusions contribute to the formation of these large bodies; for example, eight episodes of activity have been recognized in the Sierra Nevada batholith. They are formed, therefore, by the coalescence of many smaller batholiths and stocks.

Distribution of Igneous Rocks on Earth's Surface

Divergent Plate Boundaries

Most of the igneous activity on Earth is restricted to a narrow zone that is related intimately with the motions of the lithospheric plates. Indeed, the composition of the magma, the types of volcanism, and the characteristics of intrusions are governed to a large extent by plate tectonics. The magmatism at divergent plate boundaries along the crests of the oceanic rises and ridges is mostly unseen except in places where the volcanic activity occurs subaerially (e.g., Iceland, which sits on the Mid-Atlantic Ridge). Along these divergent boundaries, the erupted basalts have such a restricted compositional range that they are referred to as mid-ocean-ridge basalt (MORB). They are subalkaline tholeiites that contain olivine in the norm and less than 0.25 percent potash. The chemistry suggests that MORB was generated from a mantle that was depleted of volatile elements (e.g., lanthanum [La], cerium [Ce], sodium, and potassium) in a previous partial melting process. A wide rift valley marks the crest of most of the oceanic ridges and rises. The valley is bounded by faults created by the divergent forces and is floored in its centre by a fracture zone (a mass of rock with many small breakages). These faults and fractures are the conduits for the MORB magmas that flood the valley, build volcanoes, and produce dikes by filling the conduits. Layer 2 of the oceanic crust results from these magmatic activities. As the plates diverge, MORB becomes the ocean floor on which oceanic sediments (layer 1) are deposited. This makes MORB the most abundant rock on the surface of Earth.

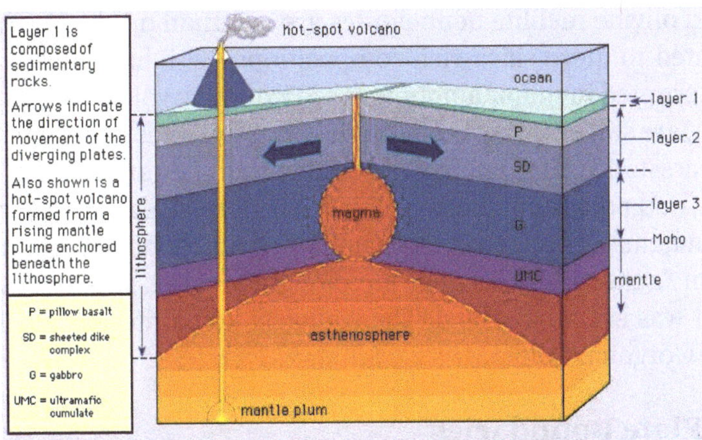

Figure: Idealized cross section of a divergent plate boundary showing the structure of the oceanic lithosphere.

Below the collection of lavas and dikes in layer 2 are found gabbro and diorite. They represent the plutonic rocks formed as a result of differentiation of the MORB magma that fed the volcanic activity along the rift. (Differentiation is the process in which more than one rock type is derived from a single parent magma.) These coarse-grained intrusives account for about 4 to 5 kilometres of layer 3, which rests on a sequence of layered ultramafic rocks. The rocks were formed by the gravitative accumulation of mafic minerals from the original MORB magma that filled a large chamber below the ridge axis. Below this layered sequence is mantle rock that is highly deformed and depleted (of elements such as lanthanum, cerium, sodium, and potassium that have been removed by repeated partial melting). Because seismic waves cannot distinguish between layered ultramafic rocks, which are not true mantle rocks, and ultramafic mantle rocks, the Moho actually is positioned between layer 3 and the layered ultramafics. The sequences consisting of layer 1 (limestone and chert sedimentary rocks), layer 2 of MORB lavas and dikes, and layer 3 of gabbro and diorite and the ultramafic rocks are known as ophiolites. Many geologists believe that ophiolites formed at oceanic ridges were emplaced by tectonic forces at convergent plate boundaries and then became exposed in highly deformed orogenic (mountain) belts. In fact, the same sequences of rocks were first reported in the Alps and were considered deep-seated intrusions. Some geologists still argue that all ophiolites were not formed at divergent plate boundaries.

Away from the axis of divergence, the composition of the volcanic rocks becomes more diverse. Most of the magmatism is related to hot spots, which are hot rising plumes of mantle rock that are anchored beneath the moving lithospheric plates. The Hawaiian Islands owe their existence to the magmatism associated with a hot spot that currently is located just southeast of the large island of Hawaii. This mantle plume not only provides magma for the eruptions at Kilauea Volcano but also is responsible for the submarine volcano named Loihi that will eventually become a new island. Most of the islands are built on a tholeiite basalt base, but the caps of the volcanoes are alkali basalts. The final episodes of volcanic activity on an island are extremely undersaturated; nephelinites and olivine melilite nephelinites are common products. The alkali basalts have differentiated to more silica-rich compositions, with hawaiites, mugearites, and trachytes being erupted in minor amounts. The two active volcanoes on Hawaii, Mauna Loa and Kilauea, are still erupting tholeiite basalts. Tholeiites on all the islands far from the ocean ridge crests are different from MORB in that they are enriched in lanthanum, cerium, sodium, and potassium. Early in Earth's history, a high-magnesium, high-temperature mafic magma called komatiite erupted from hot spots. Since most komatiites are only found in Archean regions, they are thought to be evidence for Earth being hotter than when it was initially formed. The youngest komatiite was recently discovered on the island of Gorgona, Colom.

Convergent Plate Boundaries

Igneous rocks associated with convergent plate boundaries have the greatest diversity. In this case, granite batholiths underlie the great composite volcanoes and consist of rocks ranging from basalt through andesite to dacite and rhyolite. These boundaries are destructive and consume the subducting oceanic lithosphere formed at the divergent centres. The rocks generated, however, are added on (accreted) to the continent. Oceanic trenches outline the junction of the colliding plates, but the igneous activity takes place on the overriding plate along a line at least about 100 kilometres above the subducting plate. In other words, almost no volcanism occurs between this 100-kilometre line (called the volcanic front) and the trench. The horizontal distance between the trench and the volcanic front depends on the angle of subduction; the steeper the angle, the shorter the distance. Volcanism occurs from this volcanic axis inland for a few hundred kilometres. The dominant rock constituting the composite volcanoes is andesite, but in some younger island arcs basalt tends to be more common, and in older volcanic areas dacite or rhyolite becomes prominent. Two different series of rocks are found in some volcanic chains. In Japan a tholeiitic series and a calc-alkalic series sometimes erupt from the same volcano. The former is characterized by lower magnesium, potassium, nickel, chromium, uranium, and thorium and a higher iron:magnesium ratio. Mineralogically, the tholeiitic series characteristically contains pigeonite (a low-calcium monoclinic pyroxene) in the groundmass of the basalts and andesites. The calc-alkalic series lacks pigeonite but instead has hypersthene. Most of the composite volcanoes of the Cascades Range in Oregon and Washington in the northwestern United States are characteristically calc-alkalic. In some volcanic arcs in areas farthest from the trench, a potassic series is found. In Japan the volcanoes within the Sea of Japan and farthest from the Japan Trench have alkali basalt compositions. Recent discoveries in modern convergent margins have identified igneous rocks within the oceanic trench sediments. These occur in regions where a mid-ocean ridge is being subducted. This creates higher heat flow and different types of igneous rocks, termed trondhjemite-tonalite-dacite (TTD) suites and alkaline, mafic, and felsic types.

In older areas of convergence, the composite volcanoes have been eroded, exposing the deeper plutonic granite batholiths that extend the entire length of the convergent boundaries. The batholiths are predominantly granodiorite, but gabbro through granite occur as well. It seems anomalous to find diorite, the plutonic equivalent of andesite, in low abundance since andesite is the dominant rock type of the volcanoes that were above these batholiths. Two basic types of granite have been recognized. The more common variety is located closer to the trench, has hornblende as its mafic mineral, is enriched in sodium and calcium, and has mantle chemical signatures; it is called I-type granite. The other type, called S-type granite, has muscovite and biotite and is depleted in sodium but enriched in aluminum such that corundum occurs in the norm and isotopic signatures. This suggests that such granites were formed by partial fusion of sedimentary rocks.

Flood Basalts

On the continental plates at areas away from active convergence, the magmatism is confined to rift valleys and local hot spots. The volume of magma produced is minor in comparison to that generated at oceanic rises and at convergent plate boundaries.

Flood basalts are the most common form of occurrence. They span the rock record from the Precambrian to the Neogene Period (from about 4.6 billion to 2.6 million years ago) and are found worldwide. The 1.1-billion-year-old Keweenawan flood basalts in the Lake Superior region of northern Michigan may have formed in a rift that failed. The rifting of Pangaea that began during Jurassic time (approximately 200 million to 146 million years ago) generated flood basalt eruptions all along the newly opened Atlantic Ocean. Two voluminous eruptions associated with the opening of the South Atlantic produced the Paraná basalt in Brazil and the Karoo (or Karroo) in South Africa. The Deccan basalts in India were formed in the rift valleys associated with the breakup of Gondwana during the Cretaceous Period (approximately 146 million to 65.5 million years ago). Chemically, the most abundant basalts are supersaturated tholeiites with normative quartz, but olivine tholeiites and alkali basalts also are found. Feeder dike swarms (groups consisting of many parallel dikes) and sills are common in flood basalt plateaus. Alkaline rocks, such as those found in the East African Rift System, occur as well but are less abundant. This rift system stretches southward from the Red Sea–Gulf of Aden to Lake Victoria. Undersaturated basalts are most common in these rifts. During one eruption, a magma composed mostly of sodium carbonate issued from a volcanic vent that had been erupting alkali basalts.

Other Terrestrial Occurrences

Other diverse and unusual igneous rocks are found in the stable continental areas far from plate boundaries. These include the large layered basaltic intrusions—namely, the Stillwater Complex in Montana, the Muskox intrusion in the Northwest Territories of Canada, the Bushveld Complex in South Africa, and the Skaergaard intrusion in eastern Greenland. Tholeiitic magma underwent a fractional crystallization process that deposited layers of ultramafic rocks overlain by gabbroic and anorthositic layers. The end products of this fractionation are quartz- and feldspar-bearing rocks with a peculiar texture (known as graphic intergrowth) in which quartz and feldspar are intimately intergrown with each other. These rocks are called granophyres. Such layered intrusions have some economic importance; some of them contain thick (a few metres) layers of chromite, which is the source of chromium and also platinum. Two other rare occurrences in cratonic (stable) areas of Earth's crust are the kimberlites and carbonatites. Both are of economic value because they yield diamonds and niobium, respectively. Kimberlites are mica peridotites that are found in pipes. The stable interiors of South Africa and Siberia have widespread occurrences, but these pipes also are found in North America, Australia, Brazil, and India. In North America, near Murfreesboro, Ark., individuals can pay a fee to search for diamonds in the Prairie Creek kimberlite pipe located in the Crater of Diamonds State Park. Not all kimberlites contain diamonds. When diamonds do occur, they constitute less than one part per million of the rock. Carbonatites are igneous rocks rich in carbonate (containing at least 50 percent) that commonly occur in ring complexes in association with other silica-poor rocks such as nepheline syenites. In North America, carbonatites have been found in dozens of localities in northern Ontario and western Quebec.

Extraterrestrial Occurrences

The dominant igneous rock on Earth's surface is basalt. It appears that such is also the case on Earth's close neighbours. The lunar maria are covered with basalt lava flows. These lunar basalts have a mineralogy similar to that of terrestrial basalts, but chemically they have no water, a lower amount of alkalis and alumina, and a higher iron oxide and chromium content. On the lunar highlands, plagioclase-rich rocks are most common; these include anorthosites, gabbros, troctolites (olivine-plagioclase rock), and minor basalt. It appears that basalt is common on Mars as well. The large shield volcano Olympus Mons must have been formed from eruptions of fluid basalt flows. The X-ray fluorescence analyses performed by the Vikings 1 and 2 landers showed that the rocks are basaltic. In contrast, compositions of meteorites that originated from Mars include both basalts and ultramafic rocks such as dunite, clinopyroxenite, and iherzolite. The Mars Pathfinder and Rover show that andesite may also be present, but that result is still debated. Venus apparently has volcanic features with granitic to basaltic compositions.

Rhyolites Rocks

Rhyolite is an extrusive igneous rock with a very high silica content. It is usually pink or gray in color with grains so small that they are difficult to observe without a hand lens. Rhyolite is made up of quartz, plagioclase, and sanidine, with minor amounts of hornblende and biotite. Trapped gases often produce vugs in the rock. These often contain crystals, opal, or glassy material.

Many rhyolites form from granitic magma that has partially cooled in the subsurface. When these magmas erupt, a rock with two grain sizes can form. The large crystals that formed beneath the surface are called phenocrysts, and the small crystals formed at the surface are called groundmass.

Rhyolite usually forms in continental or continent-margin volcanic eruptions where granitic magma reaches the surface. Rhyolite is rarely produced at oceanic eruptions.

Rhyolite: A pink specimen of rhyolite with numerous very tiny vugs with some evidence of flow structures. The specimen shown here is about two inches across.

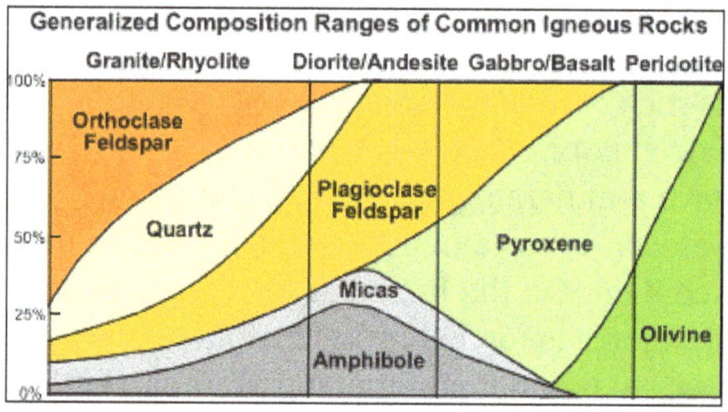

Igneous rock composition chart: This chart shows that rhyolite is typically composed of orthoclase, quartz, plagioclase, micas, and amphiboles.

Eruptions of Granitic Magma

Eruptions of granitic magma can produce rhyolite, pumice, obsidian, or tuff. These rocks have similar compositions but different cooling conditions. Explosive eruptions produce tuff or pumice. Effusive eruptions produce rhyolite or obsidian if the lava cools rapidly. These different rock types can all be found in the products of a single eruption.

Eruptions of granitic magma are rare. Since 1900 only three are known to have occurred. These were at St. Andrew Strait Volcano in Papua New Guinea, Novarupta Volcano in Alaska, and Chaiten Volcano in Chile.

Granitic magmas are rich in silica and often contain up to several percent gas by weight. As these magmas cool, the silica starts to connect into complex molecules. This gives the magma a high viscosity and causes it to move very sluggishly.

The high gas content and high viscosity of these magmas are perfect for producing an explosive eruption. The viscosity can be so high that the gas can only escape by blasting the magma from the vent.

Granitic magmas have produced some of the most explosive volcanic eruptions in Earth's history. Examples include Yellowstone in Wyoming, Long Valley in California, and Valles in New Mexico. The sites of their eruption are often marked by large calderas.

Rhyolite Porphyry: Several specimens of rhyolite porphyry, each about three inches across.

Lava Domes

Sluggish rhyolitic lava can slowly exude from a volcano and pile up around the vent. This can produce a mound-shaped structure known as a "lava dome." Some lava domes have grown to a height of several hundred meters.

Lava domes can be dangerous. As additional magma extrudes, the brittle dome can become highly fractured and unstable. The ground can also change slope as the volcano inflates and contracts. This activity can trigger a dome collapse. A dome collapse can lower the pressure on the extruding magma. This sudden lowering of pressure can result in an explosion. It can also result in a debris avalanche of material falling from the tall collapsing dome. Many pyroclastic flows and volcanic debris avalanches have been triggered by a lava dome collapse.

Lava Dome: Photo of a lava dome in the caldera of Mount St. Helens. Activity at St. Helens slowly extrudes thick lavas that gradually build domes in the caldera. This dome is composed of dacite, a rock that is intermediate in composition between rhyolite and andesite.

Rhyolite and Gemstones

Fire Opal is sometimes found filling cavities in rhyolite. This specimen of rhyolite has multiple vugs filled with gemmy transparent orange fire opal. This material can be cut into beautiful cabochons and is sometimes faceted when it is transparent or even translucent. Famous deposits of this type of fire-opal-in-rhyolite are found in Mexico.

Many gem deposits are hosted in rhyolite. These occur for a logical reason. The thick granitic lava that forms rhyolite often cools quickly while pockets of gas are still trapped inside of the lava. As the lava quickly cools, the trapped gas is unable to escape and forms cavities known as "vugs." Later, when the lava flow has cooled and hydrothermal gases or ground water move through, material can precipitate in the vugs. This is how some of the world's best deposits of red beryl, topaz, agate, jasper, and opal are formed. Gemhunters have learned this and are always on the lookout for vuggy rhyolite.

Uses of Rhyolite

Rhyolite is a rock that is rarely used in construction or manufacturing. It is often vuggy or highly fractured. Its composition is variable. When better materials are not locally available, rhyolite is sometimes used to produce crushed stone. People have also used rhyolite to manufacture stone tools, particularly scrapers, blades, and projectile points. It was probably not their material of choice, but a material used out of necessity.

Rhyolite Arrowheads: Rhyolite was often used to make stone tools and weapons when more suitable materials were not available. It has been fashioned into scrapers, hoes, axe heads, spear points, and arrowheads.

Granites and Granodiorites

Granodiorite hand specimen.

Granite and granodiorite are intrusive igneous rocks that slowly cool deep underground in magma chambers called plutons. This slow cooling process allows easily visible crystals to form. Both rocks are the product of the melting of continental rocks near subduction zones.

These rocks are both classified as granitic, because they both are rich in quartz. Granite contains mostly potassium feldspars and has a low percentage of dark iron and magnesium minerals. In contrast, granodiorite contains more plagioclase (calcium and sodium) feldspar than potassium feldspar and has more dark minerals. Thus it is a darker color than granite. Chemical and x-ray analysis of granite and granodiorite can be used to "fingerprint" these rocks, telling their exact composition and where they may have formed.

Granitic rocks are found on continents around the world near active or past plate boundaries. They formed as magma cooled many kilometers below the Earth's surface. The granitic rocks were then uplifted to the surface as the volcanic mountains above them eroded away. In California, granitic rocks form the core of the Sierra Nevada, cooled from rock melted during the subduction process that also formed the rocks of the Franciscan Complex. Granite and granodiorite are also found west of the San Andreas fault near Monterey, Pacifica, and Point Reyes, where granite from the south end of the Sierra range has been transported northward by San Andreas fault movement. Some granite was also carried to the California coast from China by humans. Chinese granite sometimes filled the holds of sailing ships on their way to San Francisco during the Gold Rush.

This type for rock is called a porphyry. The different crystal sizes are the result of different rates of cooling as the magma body moved upwards. The large crystals, called phenocrysts, are usually feldspar crystals. Feldspar is one of the first minerals to form large crystals as magma solidifies. They grew as the magma cooled very slowly deep in the magma chamber. Later, the magma with the phenocrysts moved quickly upward into cooler rock, causing more rapid cooling of the remaining molten rock to form the smaller crystals that make up the rest of the rock.

Half Dome at Yosemite National Park is composed of granodiorite.

Obsidian

Obsidian is an igneous rock that forms when molten rock material cools so rapidly that atoms are unable to arrange themselves into a crystalline structure. It is an amorphous material known as a "mineraloid." The result is a volcanic glass with a smooth uniform texture that breaks with a conchoidal fracture.

Obsidian: The specimen shown above is about two inches (five centimeters) across. The curved semi-concentric ridges are breakage marks associated with obsidian's conchoidal fracture. The rock has very sharp edges.

Formation of Obsidian

Obsidian is usually an extrusive rock - one that solidifies above Earth's surface. However, it can form in a variety of cooling environments:

- Along the edges of a lava flow (extrusive).
- Along the edges of a volcanic dome (extrusive).
- Around the edges of a sill or a dike (intrusive).
- Where lava contacts water (extrusive).
- Where lava cools while airborne (extrusive).

Types of Obsidian: The specimens shown above are from Glass Butte rockhounding site in central Oregon. It shows the diversity of obsidian types that can be found in a small geographic area. Clockwise from upper left are: double flow obsidian, rainbow obsidian, black obsidian, pumpkin obsidian, mahogany obsidian, gold sheen obsidian, and the piece in the center is gold sheen.

Mahogany obsidian: A tumble-polished specimen of "mahogany obsidian."

Color of Obsidian

Rainbow Obsidian: A baroque cabochon of iridescent "rainbow obsidian."

Black is the most common color of obsidian. However, it can also be brown, tan, or green. Rarely, obsidian can be blue, red, orange, or yellow. The colors are thought to be caused mainly by trace elements or inclusions.

Occasionally two colors of obsidian will be swirled together in a single specimen. The most common color combination is black and brown obsidian swirled together - that's called "mahogany obsidian".

As a "glass," obsidian is chemically unstable. With the passage of time, some obsidian begins to crystallize. This process does not happen at a uniform rate throughout the rock. Instead it begins at various locations within the rock. At these locations, the crystallization process forms radial clusters of white or gray cristobalite crystals within the obsidian. When cut and polished, these specimens are referred to as "snowflake obsidian".

Rarely, obsidian has an iridescent or metallic "sheen" caused by light reflecting from minute inclusions of mineral crystals, rock debris, or gas. These colored specimens are known as "rainbow obsidian," "golden obsidian," or "silver obsidian," depending upon the color of the sheen or iridescence. These specimens are very desirable for the manufacture of jewelry.

Snowflake obsidian: A tumble-polished specimen of "snowflake obsidian."

Composition of Obsidian

Most obsidians have a composition similar to rhyolite and granite. Granites and rhyolites can form from the same magma as obsidian and are often geographically associated with the obsidian.

Rarely, volcanic glasses are found with a composition similar to basalt and gabbro. These glassy rocks are named "tachylyte."

Types of Glassy Igneous Rocks

Pumice, scoria, and tachylyte are other volcanic glasses formed by rapid cooling. Pumice and scoria differ from obsidian by having abundant vesicles - cavities in the rock produced when gas bubbles were trapped in a solidifying melt. Tachylyte differs in composition - it has a composition similar to basalt and gabbro.

Occurrence of Obsidian

Obsidian outcrop: Obsidian along the edge of a lava flow in central Oregon.

Obsidian is found in many locations worldwide. It is confined to areas of geologically recent volcanic activity. Obsidian older than a few million years is rare because the glassy rock is rapidly destroyed or altered by weathering, heat, or other processes.

Significant deposits of obsidian are found in Argentina, Canada, Chile, Ecuador, Greece, Guatemala, Hungary, Iceland, Indonesia, Italy, Japan, Kenya, Mexico, New Zealand, Peru, Russia, United States, and many other locations.

In the United States it is not found east of the Mississippi River, as there is no geologically recent volcanic activity there. In the western US it is found at many locations in Arizona, California, Idaho, Nevada, New Mexico, Oregon, Washington, and Wyoming. Most obsidian used in the jewelry trade is produced in the United States.

Uses of Obsidian as a Cutting Tool

Obsidian knife blade: A knife blade manufactured from mahogany obsidian. The craftsman who made this blade had a very high skill level and was able to produce a serrated edge.

Obsidian spear point: A spear point fashioned from opaque black obsidian.

The conchoidal fracture of obsidian causes it to break into pieces with curved surfaces. This type of fracturing can produce rock fragments with very sharp edges. These sharp fragments may have prompted the first use of obsidian by people.

The first use of obsidian by people probably occurred when a sharp piece of obsidian was used as a cutting tool. People then discovered how to skillfully break the obsidian to produce cutting tools in a variety of shapes. Obsidian was used to make knives, arrowheads, spear points, scrapers, and many other weapons and tools.

Once these discoveries were made, obsidian quickly became the raw material of preference for producing almost any sharp object. The easy-to-recognize rock became one of the first targets of organized "mining." It is probably a safe bet that all natural obsidian outcrops that are known today were discovered and utilized by ancient people.

Rock and Mineral Kits: Get a rock, mineral, or fossil kit to learn more about Earth materials. The best way to learn about rocks is to have specimens available for testing and examination.

Stone Age Manufacturing and Trade

The manufacture of obsidian tools by humans dates back to the Stone Age. At some locations, tons of obsidian flakes reveal the presence of ancient "factories." Some of these sites have enough waste debris to suggest that many people labored there for decades producing a variety of obsidian objects. Making arrowheads, spear points, knife blades, and scrapers from obsidian, chert, or flint might have been the world's first "manufacturing industry."

Obsidian was so valued for these uses that ancient people mined, transported, and traded obsidian and obsidian objects over distances of up to a thousand miles. Archaeologists have been able to document the geography of this trade by matching the characteristics of obsidian in outcrops with the characteristics of obsidian in cutting tools. A study done by the Idaho National Laboratory used composition studies by X-ray fluorescence to identify the source outcrops of obsidian artifacts and map their use across the western United States.

Apache tears: "Apache Tears" is a name used for small obsidian nodules of about one inch or less that can be found in volcanic areas of the southwestern United States. Their name comes from a Native American legend. During a battle between Apaches and the U.S. Cavalry in 1870, the outnumbered Apaches, facing defeat, rode their horses over a cliff rather than allow themselves to be killed by their enemy. Upon hearing the story of the battle, the tears of their family members turned to stone when they hit the ground. Those stones are now found as the black obsidian nodules. People who do rock tumbling often polish Apache Tears. They are difficult to polish because the obsidian chips and bruises easily. Success occurs when they are cushioned during the tumbling with smaller pieces of rough or small ceramic media.

Obsidian in Modern Surgery

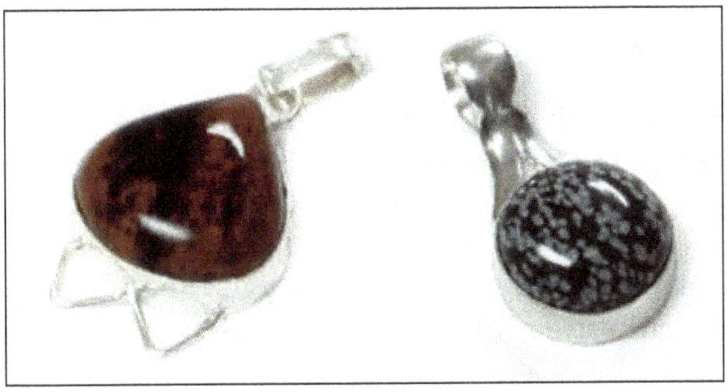

Obsidian jewelry: Mahogany obsidian and snowflake obsidian cabochons set in sterling silver pendants.

Although using a rock as a cutting tool might sound like "stone age equipment," obsidian continues to play an important role in modern surgery. Obsidian can be used to produce a cutting edge that is thinner and sharper than the best surgical steel. Today, thin blades of obsidian are placed in surgical scalpels used for some of the most precise surgery. In controlled studies, the performance of obsidian blades was equal to or superior to the performance of surgical steel.

Uses of Obsidian in Jewelry

Obsidian for opal triplets: A thin piece of obsidian is often used as a "backing" material for opal doublets and triplets. The black obsidian adds stability to the opal and provides a dark background color that contrasts with the opal's fire.

Obsidian is a popular gemstone. It is often cut into beads and cabochons or used to manufacture tumbled stones. Obsidian is sometimes faceted and polished into highly reflective beads. Some transparent specimens are faceted to produce interesting gems.

The use of obsidian in jewelry can be limited by its durability. It has a hardness of about 5.5 which makes it easy to scratch. It also lacks toughness and is easily broken or chipped upon impact. These durability concerns make obsidian an inappropriate stone for rings and bracelets. It is best suited for use in low-impact pieces such as earrings, brooches, and pendants.

Obsidian is also used in making opal doublets and opal triplets. Thin slices or chips of opal are glued to a thin slice of obsidian to make a composite stone. The black obsidian provides an inexpensive and color-contrasting background that makes opal's colorful fire much more obvious. It also adds mass and stability to the opal that facilitates cutting it into a gem.

Other uses of Obsidian

Freshly broken pieces of obsidian have a very high luster. Ancient people noticed that they could see a reflection in obsidian and used it as a mirror. Later, pieces of obsidian were ground flat and highly polished to improve their reflective abilities.

Obsidian's hardness of 5.5 makes it relatively easy to carve. Artists have used obsidian to make masks, small sculptures, and figurines for thousands of years.

Carbonatites

Carbonatites are commonly defined as magmatic rocks with high modal abundance of carbonate minerals (>50 wt%) and geochemistry typified by high abundances of Sr, Ba, P and the light rare-earth elements (LREE). They have been subdivided on the basis of their dominant modal carbonate mineral, such as calcite or dolomite carbonatites and on their corresponding major element geochemistry with Mg Ca, Fe- and REE-carbonatite:

- Calcite-carbonatite - Where the main carbonate is calcite. If the rock is coarse-grained it may be called sövite; if medium to fine-grained, alvikite.
- Dolomite-carbonatite - Where the main carbonate is dolomite. This may also be called beforsite.
- Ferrocarbonatite - Where the main carbonate is iron-rich.
- Natrocarbonatite - Essentially composed of sodium, potassium, and calcium carbonates.

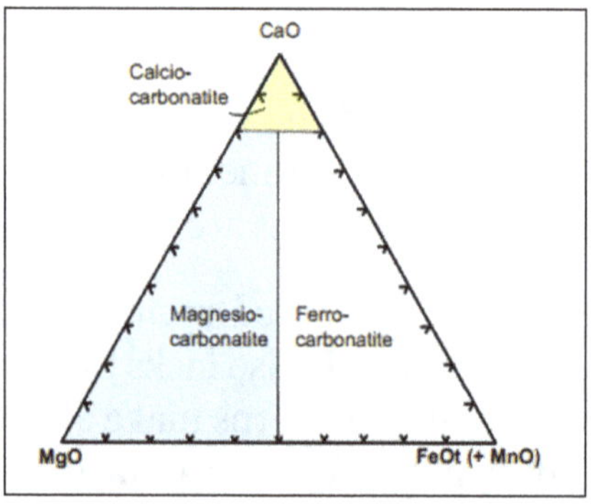

Carbonatite classification diagram.

In parallel, a process-related classification would divide them into two groups: primary carbonatites and Carbothermal residua. In this scheme, primary carbonatites can be further divided into groups of magmatic carbonatites associated with nephelinite, melilitite, kimberlite, and specific mantle-derived silicate magmas, formed by partial melting, whereas carbothermal residua carbonatites form as low-temperature fluids rich in CO_2, H_2O, and fluorine.

Carbonatites characteristically occur in close association with alkalic silicate igneous rocks, either in individual complexes or in a regional association within magmatic provinces. Associated silicates are usually alkaline ultramafic rocks: pyroxenites or nephelinites (and ijolites) generally, but may also include more evolved types such as phonolites and nepheline syenites.

The three main theories for the origin of carbonatites are essentially:

1. Residual melts of fractionated carbonated nephelinite or melilitite.
2. Immiscible melt fractions of CO_2-saturated silicate melts.
3. Primary mantle melts generated through partial melting of CO_2-bearing peridotite.

Combinations of these three theories are also popular; for example carbonatite liquids generated by deep melting of carbonated eclogite in the upper mantle infiltrate overlying peridotite to produce silica under-saturated carbonate-bearing melts, which then penetrate the crust and evolve or un-mix.

Carbonatites have also been considered to be generated in the lithospheric mantle as partial melts rising rapidly above a hot ascending mantle plume. If these mantle carbonate melts stall, for example owing to thermal death, they generate carbonate-melt metasomatism in the mantle. As the much hotter center of the plume approaches, melting is induced in the metasomatic horizon and results in generation of the carbonatite melts that are observed on the surface. Although the plume model is quite attractive, recent recognition of strong and repeated lithospheric controls in the compilation of global carbonatite ages are thought to argue against a direct connection to mantle plumes.

The atomic structures of carbonate melts have been little studied in comparison to the structure of silicate melts, but are fundamental in controlling their physical and chemical behavior in natural systems. Carbonate melts are ionic liquids consisting of carbonate CO_3^{2-} molecular anions and metal cations that interact principally due to coulombic interactions and are thus very different from silicate melts, which have network structures characterized by polymerization. Ionic carbonate melts have been considered to be structureless with no defnite association between metal cations and carbonate molecules. However the combined evidence from phase relations of carbonates, the solubility of metals in carbonate liquids, and the spectroscopy of carbonate glasses and atomic simulations, suggests that carbonate liquids have structure at scales larger than their component molecular groups.

Honey-yellow Niocalite crystals and dark Perovskite crystals in a sövite from Oka.

Jacupiranga sövite, Brasil. Calcite (withe), Magnetite (black), olivine (green).

Ca-carbonatite (sövite) from Magnet Cove.

Pyrochlore sövite (dark) from Oka.

Sövite from da Oka.

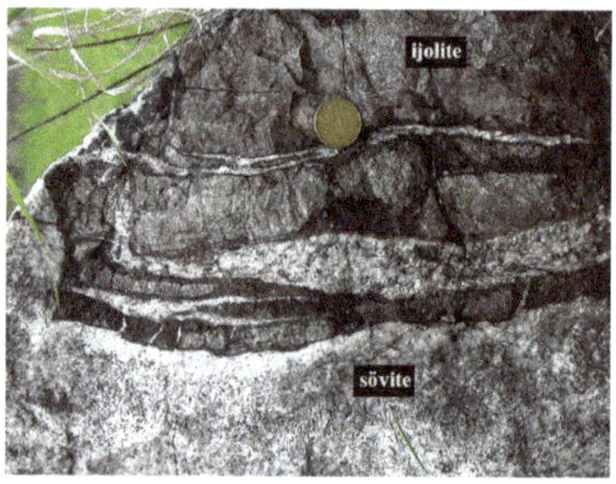

Carbonatitic veins in a ijolite. Oka.

Diorite

Diorite is the name used for a group of coarse-grained igneous rocks with a composition between that of granite and basalt. It usually occurs as large intrusions, dikes, and sills within continental crust. These often form above a convergent plate boundary where an oceanic plate subducts beneath a continental plate.

Partial melting of the oceanic plate produces a basaltic magma that rises and intrudes the granitic rock of the continental plate. There, the basaltic magma mixes with granitic magmas or melts granitic rock as it ascends through the continental plate. This produces a melt that is intermediate in composition between basalt and granite. Diorite forms if this type of melt crystallizes below the surface.

Diorite is usually composed of sodium-rich plagioclase with lesser amounts of hornblende and biotite. It usually contains little if any quartz. This makes diorite a coarse-grained rock with a contrasting mix of black and white mineral grains. Students often use this "salt and pepper" appearance as a clue to the identification of diorite.

Diorite: This specimen clearly shows the familiar "salt and pepper" appearance of diorite, produced by white plagioclase contrasting with black hornblende and biotite. This specimen is about two inches across.

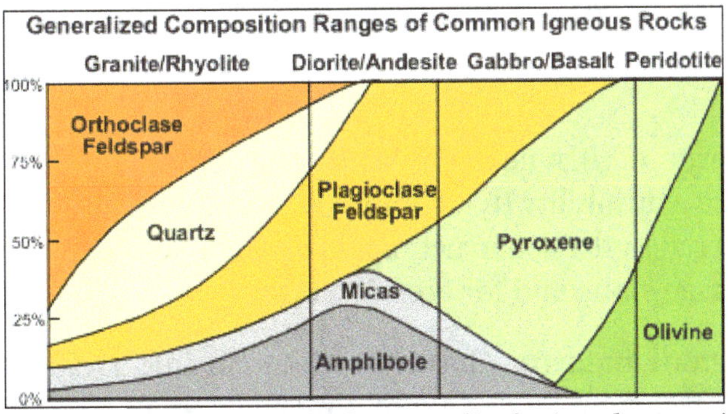

Igneous rock compositions: This chart illustrates the generalized mineral composition of igneous rocks. It shows that diorites and andesites are composed mainly of plagioclase feldspar, amphiboles, and micas; sometimes with minor amounts of orthoclase, quartz, or pyroxene.

Diorite and Andesite

Diorite and andesite are similar rocks. They have the same mineral composition and occur in the same geographic areas. The differences are in their grain sizes and their rates of cooling. Diorite crystallized slowly within the Earth. That slow cooling produced a coarse grain size. Andesite forms when a similar magma crystallizes quickly at Earth's surface. That rapid cooling produces a rock with small crystals.

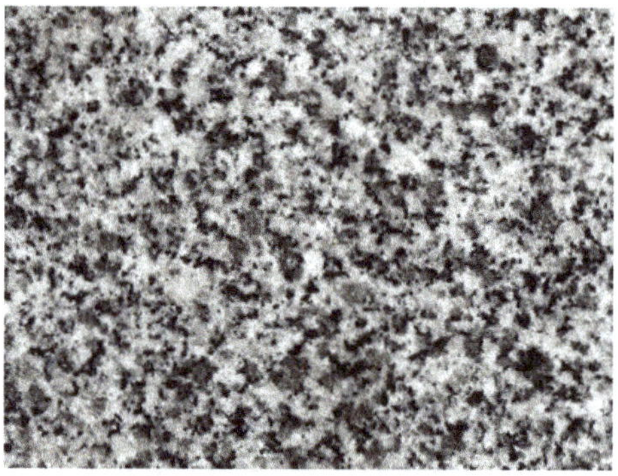

Polished diorite: A sample of diorite as it might appear in a polished countertop, facing stone, or floor tile. It would probably be marketed as "white granite" at a cabinet shop or building supply store.

Diorite Ax: Photographs of a Neolithic ax made of diorite that was found in the surroundings of Reims, France. It is in the Alexis Damour Collection at the Museum of Toulouse.

Uses of Diorite

In areas where diorite occurs near the surface, it is sometimes mined for use as a crushed stone. It has a durability that compares favorably to granite and trap rock. It is used as a base material in the construction of roads, buildings, and parking areas. It is also used as a drainage stone and for erosion control.

In the dimension stone industry, diorite is often cut into facing stone, tile, ashlars, blocking, pavers, curbing, and a variety of dimension stone products. These are used as construction stone, or polished and used as architectural stone. Diorite was used as a structural stone by the Inca and Mayan civilizations of South America and by many ancient civilizations in the Middle East.

In the dimension stone industry, diorite is sold as a "granite." The dimension stone industry uses the name "granite" for any rock with visible, interlocking grains of feldspar. This simplifies discussions with customers who do not know how to identify igneous and metamorphic rocks.

Fossil and mineral kit.

Diorite in Art

Diorite is difficult to sculpt because of its hardness, variable composition, and coarse grain size. For those reasons, it is not a favored stone of sculptors, although it was popular among ancient sculptors of the Middle East.

The most famous diorite sculpture is the Code of Hammurabi, a black diorite pillar about seven feet tall, inscribed with Babylonian laws in about 1750 BC.

Diorite has the ability to accept a bright polish, and it has occasionally been cut into cabochons or used as a gemstone. In Australia, a diorite with beautiful pink feldspar phenocrysts has been cut into cabochons and called “pink marshmallow stone.”

Diorite Cabochon: A diorite in Australia contains large, beautiful pink feldspar crystals. It is often cut into cabochons for use as a novelty gem. It has been given the name “pink marshmallow stone.”

Diorite Sculptures: The sculpture above is a diorite statue of Gudea, a Mesopotamian ruler, made in about 2090 BC. It is about 19 inches tall and is currently displayed at the

Metropolitan Museum of Art. A public domain image. The vase on the right was made in ancient Egypt from diorite with spectacular feldspar phenocrysts. It is in the collection of the Field Museum.

Andesite

Andesite is the name used for a family of fine-grained, extrusive igneous rocks that are usually light to dark gray in color. They often weather to various shades of brown, and these specimens must be broken for proper examination. Andesite is rich in plagioclase feldspar minerals and may contain biotite, pyroxene, or amphibole. Andesite usually does not contain quartz or olivine.

Andesite is typically found in lava flows produced by stratovolcanoes. Because these lavas cooled rapidly at the surface, they are generally composed of small crystals. The mineral grains are usually so small that they cannot be seen without the use of a magnifying device. Some specimens that cooled rapidly contain a significant amount of glass, while others that formed from gas-charged lavas have a vesicular or amygdaloidal texture.

Andesite: The specimen shown is about two inches (five centimeters) across and has a porphyritic texture.

Porphyritic andesite.

Origin of Andesite

Andesite and diorite are common rocks of the continental crust above subduction zones. They generally form after an oceanic plate melts during its descent into the subduction zone to produce a source of magma. Diorite is a coarse-grained igneous rock that forms when the magma did not erupt, but instead slowly crystallized within Earth's crust. Andesite is a fine-grained rock that formed when the magma erupted onto the surface and crystallized quickly.

Andesite and diorite have a composition that is intermediate between basalt and granite. This is because their parent magmas formed from the partial melting of a basaltic oceanic plate. This magma may have received a granitic contribution by melting granitic rocks as it ascended or mixed with granitic magma.

Andesite derives its name from the Andes Mountains of South America. In the Andes it occurs as lava flows interbedded with ash and tuff deposits on the steep flanks of stratovolcanoes. Andesite stratovolcanoes are found above subduction zones in Central America, Mexico, Washington, Oregon, the Aleutian Arc, Japan, Indonesia, the Philippines, the Caribbean, and New Zealand, among other locations.

Andesite can also form away from the subduction zone environment. For example, it can form at ocean ridges and oceanic hot spots from partial melting of basaltic rocks. It can also form during eruptions at continental plate interiors where deep-source magma melts continental crust or mixes with continental magmas. There are many other environments where andesite might form.

Stratovolcanoes: Pavlof Volcano (right) and Pavlof Sister Volcano (left) are a pair of symmetrical stratovolcanoes built of andesite flows and tephra on the Alaska Peninsula. Pavlof Volcano is one of the most active volcanoes in Alaska.

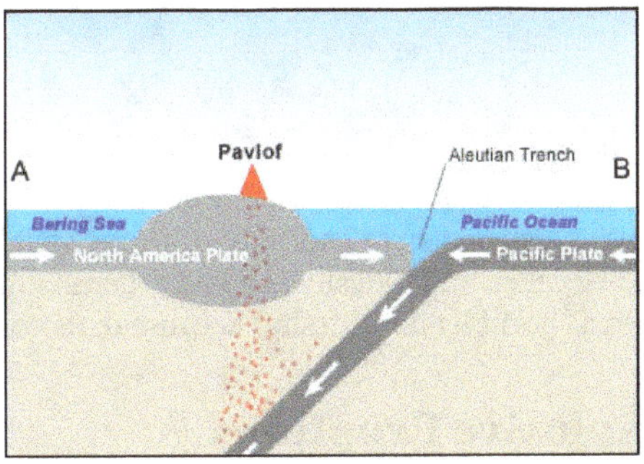

Pavlof Volcano - plate tectonics: Simplified plate tectonics cross-section showing how Pavlof Volcano is located above a subduction zone where basaltic crust of the Pacific

Plate is being partially melted at depth. The ascending magma then passes through continental crust, where it might mix with other magmas or be altered by melting rocks of different composition.

Andesite Porphyry

Occasionally, andesites contain large, visible grains of plagioclase, amphibole, or pyroxene. These large crystals are known as "phenocrysts." They begin forming when a magma, which is cooling at depth, approaches the crystallization temperature of some of its minerals. These high-crystallization-temperature minerals begin forming below the surface and grow to visible sizes before the magma erupts.

When the magma erupts onto the Earth's surface, the rest of the melt crystallizes quickly. This produces a rock with two different crystal sizes: large crystals that formed slowly at depth (known as "phenocrysts"), and small crystals that formed quickly at the surface (known as "groundmass"). "Andesite porphyry" is the name used for these rocks with two crystal sizes.

Andesite outcrop: Close view of an andesite lava flow at Brokeoff Volcano in California.

Hornblende Andesite Porphyry: A specimen of andesite with large visible hornblende phenocrysts. This type of rock could be called an "andesite porphyry" because of its texture. It could also be called a "hornblende andesite" because of its composition.

Dissolved Gas and Explosive Eruptions

Some magmas that produce volcanic eruptions above subduction zones contain enormous amounts of dissolved gas. These magmas can contain several percent dissolved gas by weight. This gas can have several origins, examples of which include the following:

- Water vapor produced when ocean-floor sediments on an oceanic plate are heated in a subduction zone.
- Water vapor produced when hydrous minerals dehydrate in the heat of a subduction zone.
- Carbon dioxide produced when rising magma encounters carbonate rocks, such as limestone, marble, or dolomite.
- Water vapor produced when a rising magma chamber encounters groundwater.

At depth, these gases can be dissolved in the magma like carbon dioxide dissolved in a can of cold beer. If that can of beer is shaken and suddenly depressurized by opening the can, the gas and the beer will erupt from the opening. A volcano behaves in a similar manner. A rising magma chamber instantly depressurized by a landslide, faulting, or other event can produce a similar but much larger explosive eruption.

Many volcanic plumes and ash eruptions occur when gas-charged andesitic magmas erupt. The gas pressure that causes the eruption blows large amounts of tiny rock and magma particles into the atmosphere. These particles can be blown high into the atmosphere and carried long distances by the wind. They often cause problems for aircraft operating downwind from the volcano.

Catastrophic eruptions like Mount St. Helens, Pinatubo, Redoubt, and Novarupta were produced by andesitic magmas with enormous amounts of dissolved gas under high pressure. It is difficult to imagine how a magma can contain enough dissolved gas to produce one of these eruptions.

Andesite Flow: One of numerous massive andesite flows from the Zarembo Island area of southeastern Alaska. They are gray pyroxene and feldspar porphyrys that weather to maroon or green.

The Elusive Definition of Andesite

The formal definition of andesite is problematic. Many authors have classified igneous rocks based upon their chemical and mineralogical compositions. However, none of these classifications are in perfect agreement.

For a fine-grained rock like andesite, these classifications are impossible to use precisely when in the field or the classroom. They require chemical or mineralogical analyses that are usually not available, affordable, or practical.

If you examine a rock that appears to be andesite, but you are not confident that it meets the mineralogical or chemical classification of andesite, you can properly call it an "andesitoid" rock. That means that while the rock looks like andesite, a microscopic examination or chemical testing might prove you wrong.

Diabase

Diabase are also called Dolerite, fine- to medium-grained, dark gray to black intrusive igneous rock. It is extremely hard and tough and is commonly quarried for crushed stone, under the name of trap. Although not popular, it makes an excellent monumental stone and is one of the dark-coloured rocks commercially known as black granite. Diabase is widespread and occurs in dikes (tabular bodies inserted in fissures), sills (tabular bodies inserted while molten between other rocks), and other relatively small, shallow bodies. Chemically and mineralogically, diabase closely resembles the volcanic rock basalt, but it is somewhat coarser and contains glass. With increase in grain size, diabase may pass into gabbro.

About one-third to two-thirds of the rock is calcium-rich plagioclase feldspar; the remainder is mostly pyroxene or hornblende. In diabase, poorly formed pyroxene crystals wrap around or mold against long, rectangular plagioclase crystals to give it the characteristic texture known as diabasic or ophitic. The larger pyroxene grains may completely enclose plagioclase; but as the quantity of the latter increases, pyroxene appears more interstitial.

Certain flat tabular masses (thick sheets or sills) of diabase, such as that forming the Palisades along the Hudson River near New York City, show concentrations of heavy minerals (as olivine or pyroxene) in their lower portions. These concentrations are commonly believed to have developed by the settling of early formed crystals in molten diabase.

Diabase may show varying degrees of alteration: plagioclase is converted to sassurite; pyroxene to hornblende, actinolite, or chlorite; and olivine to serpentine and magnetite. In British usage, such altered rock is called diabase. Some diabase masses have been subdivided by systematic fractures into rectangular blocks. Subsequent alteration and weathering along these fractures have disintegrated and rounded off block corners and edges (spheroidal weathering), leaving regularly spaced, spherelike masses of fresh diabase enveloped by shells of progressively more altered and disintegrated material.

Tuff

Tuff is an igneous rock that forms from the products of an explosive volcanic eruption. In these eruptions, the volcano blasts rock, ash, magma and other materials from its vent. This ejecta travels through the air and falls back to Earth in the area surrounding the volcano. If the ejected material is compacted and cemented into a rock, that rock will be called "tuff."

Tuff is usually thickest near the volcanic vent and decreases in thickness with distance from the volcano. Instead of being a "layer," a tuff is usually a "lens-shaped" deposit. Tuff can also be thickest on the downwind side of the vent or on the side of the vent where the blast was directed.

Some tuff deposits are hundreds of meters thick and have a total eruptive volume of many cubic miles. That enormous thickness can be from a single eruptive blast or, more commonly, from successive surges of a single eruption - or eruptions that were separated by long periods of time.

Tuff: An igneous rock that contains the debris from an explosive volcanic eruption.

Beryllium tuff: A specimen of beryllium tuff from the Spor Mountain area of Utah.

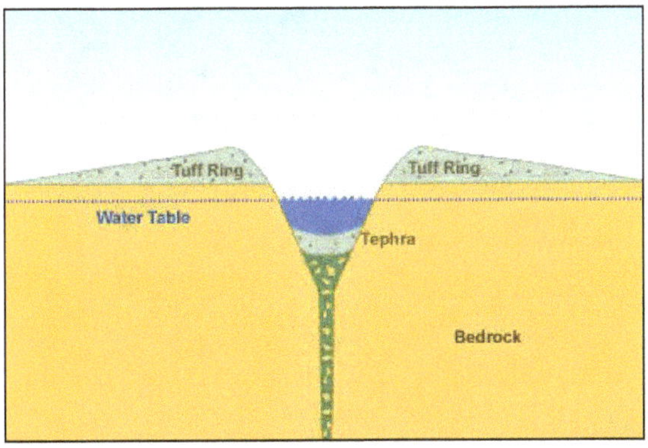

Tuff ring: Drawing of a tuff ring surrounding a shallow, water-filled crater. The tuff ring is formed from materials that were ejected by the volcanic blast and fell back to Earth in the area surrounding the crater.

Tuff Rings

A "tuff ring" is a small volcanic cone of low relief that surrounds a shallow crater. These craters, known as maars, are formed by explosions caused by hot magma coming in contact with cold groundwater. The explosion blasts fragments of bedrock, tephra, and ash from the crater. The tuff ring forms as these ejected materials fall back to Earth. Tuff rings range in size from several hundred meters across to several thousand meters. They are typically less than a few hundred meters in height and have a very gentle slope of less than ten degrees.

Tuff: Close-up of a piece of tuff exposed at Hole-in-the-Wall, Mojave National Preserve, California. This specimen clearly displays the diversity of materials that compose a tuff.

Welded Tuff

Sometimes the ejecta is hot enough when it lands that the particles are soft and sticky. These materials "weld" together upon impact or upon compaction. The rock formed from this hot ejecta is known as a "welded tuff" - because the ejected particles are welded together. Some deposits might contain welded tuff near the vent and unwelded tuff at a distance where smaller, cooler particles fell to the ground.

Ettringer tuff: Close-up of a specimen of Ettringer Tuff showing a variety of rock fragments and tephra in a matrix of volcanic ash.

Many Types of Tuff

"Tuff" is a name that is used for a broad range of materials. The only requirement is that the materials are ejecta produced by a volcanic eruption. Tuff can contain fragments of dust-size particles to boulder-size particles and be composed of many different types of material.

Mount St. Helens tephra: Photograph of an outcrop of stratified tuff that formed from tephra produced by pre-1980 eruptions at Mount St. Helens, Washington.

Many tuff deposits contain fragments of bedrock that are unrelated to volcanic activity. These materials are involved when the volcanic explosion occurs below the ground. The subsurface explosion crushes the overlying bedrock and launches it into the air mixed with tephra and volcanic ash produced from the magma source below.

Different volcanoes are supplied with magma of different compositions. Many tuff deposits form from magma with a rhyolitic composition, but andesitic, basaltic, and other types of magma might contribute to the tuff.

Tuff also varies by particle size. Near the vent, a tuff might consist mainly of large blocks of material in a volcanic ash matrix. With distance from the vent, the clasts will be smaller in size. At the edges of the rock unit, the tuff might be mainly composed of very fine ash.

Fire Opal

Many people confuse "fire opal" with "precious opal". The three basic types of opal:

1. Fire Opal is a variety of opal that has a bright yellow, bright orange or bright red background color. The stones in the first photo on this page are fire opal. They receive their name from their fiery background color.
2. Precious Opal is a name given to any opal that exhibits "play-of-color", a flashing

display of spectral colors when the opal is "played" under a light source. The second photo on this page shows many varieties of precious opal.

3. Common Opal is a variety of opal that does not display "play-of-color" and does not have a background color that would make it a fire opal.

The name "fire opal" is all about the bodycolor - fiery yellow, fiery orange or fiery red. It's not about "flash".

Mexican Fire Opal: Cabochons cut from fire opal found in Mexico. They all have a bright red, orange or yellow background color.

Faceted Fire Opal: Some beautiful faceted fire opals from Mexico. The red stone measures 9 x 6 mm, the orange stone measures 7 x 6 mm, and the yellow stone measures 8 x 6 mm.

Precious Fire Opal

A few specimens of fire opal exhibit "play-of-color". These specimens have both the fiery background color of a fire opal and the play-of-color of a precious opal. Some people call these special stones "precious fire opal".

The play-of-color can be tiny bright flashes of color when the opal is played under light, or a billowing glow contained within the stone. The accompanying photo is a precious fire opal cut from material found in Ethiopia. It exhibits a neon green play-of-color from one direction and a neon violet play-of-color from another direction. From a few directions you can see a little of both.

In the past decade, Ethiopia has become a steady producer of beautiful opal. Much of it is yellow to orange semitransparent opals with play-of-color.

Faceted Fire Opal: These three stones show the color range of "fire opal," a name given to specimens of opal with a fiery background color. The orange and yellow stones have a sleepy translucence, while the red stone is semitranslucent, almost opaque.

The orange stone is about 7 x 9 millimeters in size and was mined in Oregon. Fire opal of this color is sometimes called "tangerine opal".

The red stone is about 8 x 10 millimeters in size and was mined in Mexico. Fire opal of this color is often called "cherry opal".

The yellow stone is about 9 millimeters across and was mined in Nevada. Fire opal of this color is sometimes called "lemon opal".

Ways in which Fire Opals Cut

Fire opals are cut in a variety of ways. Some are cut as faceted stones, others are cut as cabochons. The cutter decides how he/she thinks the stone will be most attractive. There is no rule for cutting fire opal.

Transparent fire opals are often faceted so that they can be illuminated by incident light. If they have a spectacular play-of-color, they might be cut into a cabochon like most precious opal. If the play-of-color is minor, it might be cut into a faceted stone with a little surprise of flash.

Translucent stones are often cut into cabochons, but it is not unusual to see a translucent to nearly opaque fire opal with an attractive color cut into a pretty faceted stone. The three faceted stones in the photo above are wonderful examples of translucent stones that have been faceted.

Cantera Opal: Much of the fire opal found in Mexico is in a rhyolite host rock. It is difficult to remove the opal from the rhyolite without damage. So, some cutters simply cut a cab of rhyolite with a fire opal window in the dome. Cabs with this style of cutting are known as "cantera opal". Many people enjoy them.

Fire Opal Value

The value of a fire opal is based upon the desirability and uniformity of its color, with yellow being on the low end of value and red being on the high end.

Transparent stones are preferred over translucent stones. The best fire opal typically sells for prices that are much lower than the best precious opal; however, fire opal specimens with exceptional color will sell for higher prices than some specimens of precious opal with less impressive play-of-color.

Durability of Fire Opal

Fire opal has a Mohs hardness of 5.5 to 6, which is soft enough that it can be scratched by many objects that it might encounter during everyday wear. Fire opal also has a low tenacity, which means that it can easily be chipped or broken.

Fire opal is best used in jewelry such as earrings, pins, and pendants that usually are not subjected to rough wear. If a fire opal is set into a ring, a setting especially designed to protect the stone from abrasion and impact is recommended.

Fire Opal Localities

Mexico has been the world's primary source of fire opal for nearly 100 years. The Mexican deposits produce significant amounts of transparent to translucent, bright orange to orange-red material. Some of the transparent material is faceted, mounted in commercial jewelry, and described as "tangerine opal" because of its color.

In the 1990s, Ethiopia became an important source of opal. Much of the Ethiopian opal is fire opal and precious fire opal. Smaller amounts of fire opal are produced in Australia, Brazil, Honduras, Guatemala, Nevada, and Oregon.

Gabbro

Gabbro is a coarse-grained and usually dark-colored igneous rock. It is an intrusive rock. It means that it formed as magma cooled slowly in the crust. Igneous rocks with similar composition are basalt (extrusive equivalent of gabbro) and diabase (the same rock type could be named dolerite or microgabbro instead).

Gabbro with large augite phenocrysts embedded in white plagioclase. These two are the principal minerals in this rock.

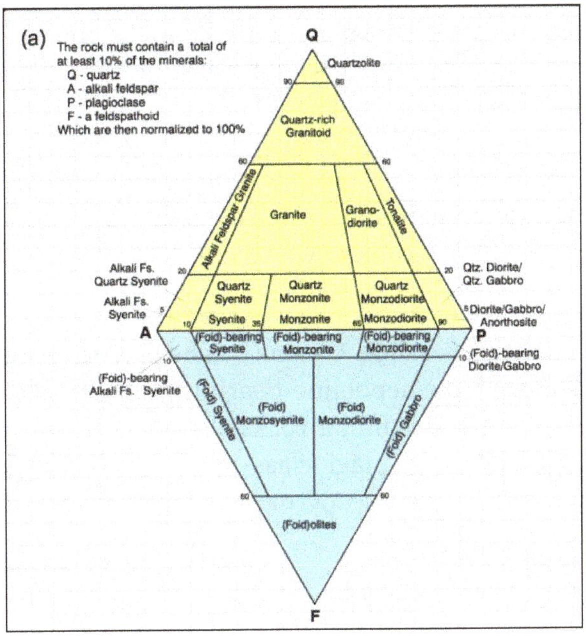

The most important mineral groups that make up this rock type are plagioclase and pyroxene. Plagioclase usually predominates over pyroxene. Plagioclase is sodium-calcium feldspar. It contains more calcium than sodium in gabbro. If there is more sodium in the plagioclase, then the rock type is named diorite. Diorite is usually lighter in color and contains more amphiboles than pyroxenes.

"Gabbro" in the strict sense of the term is an intrusive rock that is chiefly composed of monoclinic pyroxene (clinopyroxene, abbreviated Cpx) and plagioclase. If more than 5% of Cpx is replaced with orthorhombic pyroxene (orthopyroxene, abbreviated Opx), the rock is named gabbronorite. If more than 95% of the pyroxene is Opx, then we have norite. These rocks are all collectively called gabbroic rocks. Sometimes the term "gabbro" is used loosely to include them all. Rocks that contain more than 90% pyroxene is pyroxenite. Gabbroic rocks may contain small amounts of quartz (up to 5%). If there is more quartz present, the rock must be named quartz gabbro. If quartz makes up more than 20%, then the rock is one of granitoids.

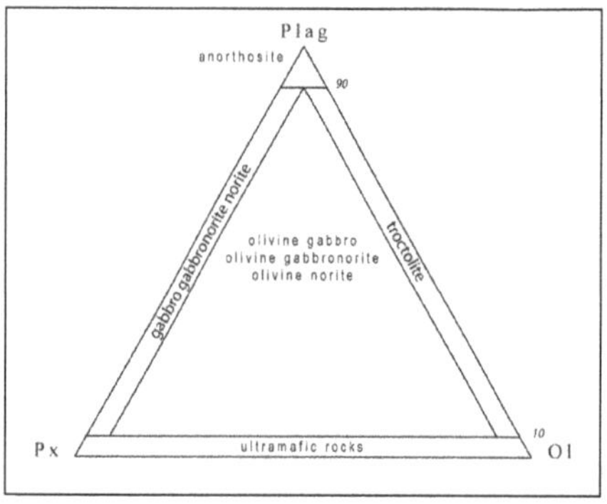

The classification scheme of olivine-bearing (>10%) gabbroic rocks. "Olivine" should be added to the rock name if it contains more than 10% of it.

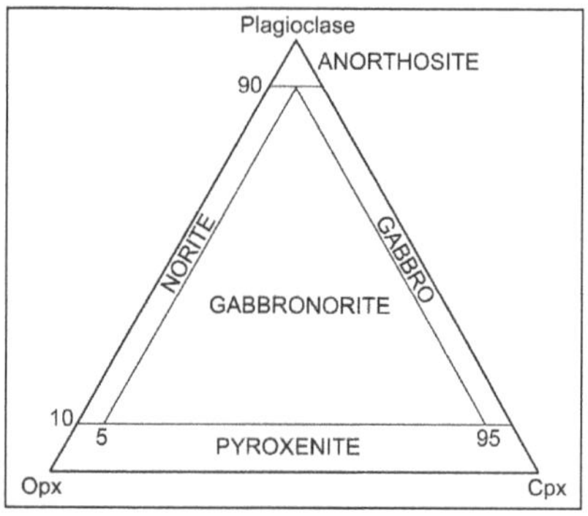

The classification scheme for gabbroic rocks without taking into account olivine and feldspathoids. Cpx is clinopyroxene, Opx is orthopyroxene, Plag is plagioclase.

Rocks that contain less than 10% plagioclase are ultramafic (pyroxenite if pyroxenes are the dominant minerals). Pyroxene is almost exclusively clinopyroxene and norite is conversely rich in orthopyroxene. Anorthosite is almost monomineralic plagioclase-rock.

There are even more varieties possible. Olivine can be present or feldspathoids just like in basalt. Hence, rock names like nepheline-bearing gabbro or olivine norite are possible. Anorthosite and troctolite are similar rocks also. Anorthosite is almost monomineralic rock — more than 90% of the plagioclase-pyroxene pair is plagioclase. Troctolite contains mostly olivine and plagioclase, pyroxene again forms no more than 10% of the composition.

It is often mentioned that the term "gabbro" was brought into geological terminology by a German geologist Leopold von Buch. It is probably not correct. The term was used by Italian geologist Tozzetti in 1768. The term itself comes from Italy. Von Buch redefined the term in 1810 as a rock composed of diallage and feldspar or saussurite. Saussurite is a mixture of epidote, albite, and other alteration minerals. Saussuritized version can be named metagabbro nowadays because its original composition has clearly been altered by metamorphism. Diallage is not defined as a mineral now, but it is monoclinic pyroxene anyway (either diopside or augite). So we can say that von Buch used the term roughly as we use it today.

Gabbros form in the crust. This is the magma that did not break to the surface to cool as a basalt. These rocks are not as widespread as granitoids, but they are definitely not rare. Gabbroic rocks are usually equigranular (composed of similarly sized grains) mixtures of black, brown or greenish pyroxene and white, gray, or greenish plagioclase. Pyroxenite and other ultramafic rocks are darker- and diorite is lighter-colored.

A hand sample from the Isle of Skye in Scotland (Black Cuillin Mountains). Width of sample 55 mm.

Olivine gabbro. Olivine (weathered orange iddingsite in this sample) may be an important constituent in these rocks. Width of sample 12 cm.

Gabbro pegmatite from Cyprus (the Troodos ophiolite).

A sample from Tangen, Norway. Width of sample 10 cm.

Pumice

Pumice: This specimen shows the frothy vesicular texture of pumice. It has a specific gravity of less than one and will float on water. It is about five centimeters (two inches) across.

Pumice is a light-colored, extremely porous igneous rock that forms during explosive volcanic eruptions. It is used as aggregate in lightweight concrete, as landscaping aggregate, and as an abrasive in a variety of industrial and consumer products. Many specimens have a high enough porosity that they can float on water until they slowly become waterlogged.

Pumice at Mount St. Helens: A pyroclastic flow will sometimes contain large pieces of pumice. This photograph shows a USGS scientist examining blocks of pumice at the toe of a pyroclastic flow at Mount St. Helens.

Pumice quarry: Photograph of stratified pumice deposit produced by pyroclastic flows at Mount St. Helens, Washington.

Formation of Pumice

The pore spaces (known as vesicles) in pumice are a clue to how it forms. The vesicles are actually gas bubbles that were trapped in the rock during the rapid cooling of a gas-rich frothy magma. The material cools so quickly that atoms in the melt are not able to arrange themselves into a crystalline structure. Thus, pumice is an amorphous volcanic glass known as a "mineraloid."

Some magmas contain several percent dissolved gas by weight while they are under pressure. Stop for a moment and think about that. Gas weighs very little at Earth's surface, but these magmas under pressure can contain several percent gas by weight held in solution.

This is similar to the large amount of dissolved carbon dioxide in a sealed bottle of carbonated beverage such as beer or soda. If you shake the container, then immediately open the bottle, the sudden release of pressure allows the gas to come out of solution, and the beverage erupts from the container in a frothy mess.

A rising body of magma, supercharged with dissolved gas under pressure, behaves in a similar way. As the magma breaks through Earth's surface, the sudden pressure drop causes the gas to come out of solution. This is what produces the enormous rush of high-pressure gas from the vent.

This rush of gas from the vent shreds the magma and blows it out as a molten froth. The froth rapidly solidifies as it flies through the air and falls back to Earth as pieces of pumice. The largest volcanic eruptions can eject many cubic kilometers of material. This material can range in size from tiny dust particles to large blocks of pumice the size of a house.

Large eruptions can blanket the landscape around the volcano with over 100 meters of pumice and launch dust and ash high into the atmosphere.

Gas and Pumice at the Pinatubo Eruption

Pinatubo eruption: The explosive eruption of Mount Pinatubo in the Philippines on June 12, 1991 ejected more than five cubic kilometers of material and was rated as a VEI 5 eruption on the volcanic explosivity index. Much of that material was pumice lapilli that blanketed the landscape around the volcano.

Pinatubo pumice: Dacitic pumice fragments erupted by Mount Pinatubo, Philippines, during an enormous eruption on 15 June 1991.

The second most powerful volcanic eruption of the 20th century was at Mount Pinatubo in 1991. The description below explains how enormous volumes of dissolved gas powered the eruption and how a cubic mile of ash and pumice lapilli was blasted from the volcano.

"From June 7 to 12, the first magma reached the surface of Mount Pinatubo. Because it had lost most of the gas contained in it on the way to the surface, the magma oozed out to form a lava dome but did not cause an explosive eruption. However, on June 12, millions of cubic yards of gas-charged magma reached the surface and exploded in the reawakening volcano's first spectacular eruption.

When even more highly gas charged magma reached Pinatubo's surface on June 15, the volcano exploded in a cataclysmic eruption that ejected more than 1 cubic mile of material. A blanket of volcanic ash and pumice lapilli blanketed the countryside.

Huge avalanches of searing hot ash, gas, and pumice roared down the flanks of Mount Pinatubo, filling once-deep valleys with fresh volcanic deposits as much as 660 feet thick. The eruption removed so much magma and rock from below the volcano that the summit collapsed to form a large volcanic depression 1.6 miles across."

Mount Mazama Eruption (Crater Lake)

"The cataclysmic eruption of Mount Mazama 7,700 years ago started from a single vent on the northeast side of the volcano as a towering column of pumice and ash that reached some 30 miles high. Winds carried the ash across much of the Pacific Northwest and parts of southern Canada. So much magma erupted that the volcano began to collapse in on itself. As the summit collapsed, circular cracks opened up around the peak. More magma erupted through these cracks to race down the slopes as pyroclastic flows. Deposits from these flows partially filled the valleys around Mount Mazama with up to 300 feet of pumice and ash. As more magma was erupted, the collapse progressed until the dust settled to reveal a volcanic depression, called a caldera, 5 miles in diameter and one mile deep."

Pumice raft: View of a pumice raft from a boat. Waves can be seen moving under the pumice. The rafts can float for years until all of the pumice becomes waterlogged and sinks or it is dissipated by waves and wind.

Pumice raft: A "raft" of lightweight pumice floating on the surface of the South Pacific after an eruption in the Tonga Islands.

Composition of Pumice

Most pumice erupts from magmas that are highly charged with gas and have a rhyolitic composition. Rarely, pumice can erupt from gas-charged magmas of basaltic or andesitic composition.

Pantheon: Some of the concrete used to construct the Pantheon by the Romans in 126 AD was lightweight material made with pumice aggregate.

Pumice has a very Low Specific Gravity

The abundant vesicles in pumice and the thin walls between them give the rock a very low specific gravity. It typically has a specific gravity of less than one, giving the rock an ability to float on water.

Pumice products: A variety of health and beauty products that contain pumice. They include the famous "Lava Soap" that cleans dirty hands with tiny pieces of pumice abrasive, a foot scrub cream that works as an exfoliant to smooth "sandal feet," two pumice stones, and a sponge with embedded pumice abrasive.

Large amounts of pumice produced by some island and subsea eruptions will float on the surface and be pushed about by the winds. The pumice can float for long periods of time - sometimes years - before it finally becomes waterlogged and sinks. Large masses of floating pumice are known as "pumice rafts." They are large enough to be tracked by satellites and are a hazard to ships that sail through them.

Uses of Pumice

The largest use of pumice in the United States is the production of lightweight concrete blocks and other lightweight concrete products. When this concrete is mixed, the vesicles remain partially filled with air. That reduces the weight of the block. Lighter blocks can reduce the structural steel requirements of a building or reduce the foundation requirements. The trapped air also gives the blocks a greater insulating value.

The second most common use of pumice is in landscaping and horticulture. The pumice is used as a decorative ground cover in landscaping and planters. It is used as drainage rock and soil conditioner in plantings. Pumice and scoria are also popular rocks for use as substrates in hydroponic gardening.

Pumice has many other uses. Together these account for less than a few percent of consumption in the United States, but these are the products that most people think of when they hear the word "pumice."

Lots of people have found small pumice pebbles in the pockets of brand new "stone washed jeans," and almost everyone has seen the famous "Lava Soap" that uses pumice as an abrasive. Below we list these and some of the other minor uses of pumice:

- An abrasive in conditioning "stone washed" denim.
- An abrasive in bar and liquid soaps such as "lava soap".
- An abrasive in pencil erasers.
- An abrasive in skin exfoliating products.
- A fine abrasive used for polishing.
- A traction material on snow-covered roads.
- A traction enhancer in tire rubber.
- An absorbent in cat litter.
- A fine-grained filter media.
- A lightweight filler for pottery clay.

Pumice and Pumicite Production

Pumice is produced in two forms: rock pumice and pumicite. "Pumicite" is a name given to very fine-grained pumice (less than 4 millimeters in diameter down to submillimeter sizes). The word can be used synonymously with "volcanic ash." It is mined from volcanic ash deposits, or it can be produced by crushing rock pumice.

About 500,000 metric tons of pumice and pumicite were mined in the United States in 2011. The total value of this pumice was about $11,200,000, or an average of about $23 per ton at the mine. The producing states were, in order of decreasing production:

- Oregon
- Nevada
- Idaho
- Arizona
- California
- New Mexico
- Kansas.

Pumice Reticulite: Reticulite is a basaltic pumice in which all of the bubbles have burst, leaving a honeycomb structure.

Imported Pumice and Substitutes

All of the pumice production in the United States occurs west of the Mississippi River. In 2011, most of the pumice for consumption in the eastern United States was imported from Greece.

In the eastern United States, expanded aggregate, produced by heating specific types of shale under controlled conditions, is used as a substitute for pumice in lightweight aggregate, horticultural, and landscaping applications.

Magma and Magma Formation

Magmas can vary widely in composition, but in general they are made up of only eight elements; in order of importance: oxygen, silicon, aluminum, iron, calcium, sodium, magnesium, and potassium. Oxygen, the most abundant element in magma, comprises a little less than half the total, followed by silicon at just over one-quarter. The remaining elements make up the other one-quarter. Magmas derived from crustal material are dominated by oxygen, silicon, aluminum, sodium, and potassium.

The composition of magma depends on the rock it was formed from (by melting), and the conditions of that melting. Magmas derived from the mantle have higher levels of iron, magnesium, and calcium, but they are still likely to be dominated by oxygen and silicon. All magmas have varying proportions of elements such as hydrogen, carbon, and sulphur, which are converted into gases like water vapour, carbon dioxide, and hydrogen sulphide as the magma cools.

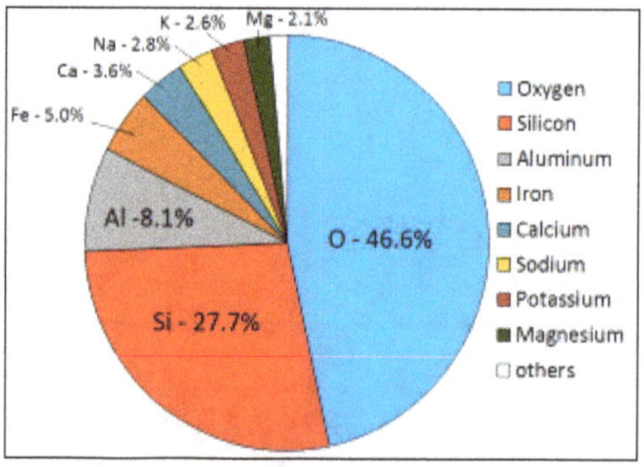

Figure: Average elemental proportions in Earth's crust, which is close to the average composition of magmas within the crust [SE].

Virtually all of the igneous rocks that we see on Earth are derived from magmas that formed from partial melting of existing rock, either in the upper mantle or the crust. Partial melting is what happens when only some parts of a rock melt; it takes place because rocks are not pure materials. Most rocks are made up of several minerals, each of which has a different melting temperature. The wax in a candle is a pure material. If you put some wax into a warm oven (50 °C will do as the melting temperature of most wax is about 40 °C) and leave it there for a while, it will soon start to melt. That's complete melting, not partial melting. If instead you took a mixture of wax, plastic, aluminum, and glass and put it into the same warm oven, the wax would soon start to melt, but the plastic, aluminum, and glass would not melt. That's partial melting and the result would be solid plastic, aluminum, and glass surrounded by liquid wax. If we heat the oven up to around 120° C, the plastic would melt too and mix with the liquid wax, but the aluminum and glass would remain solid. Again this is partial melting. If we separated the wax/plastic "magma" from the other components and let it cool, it would eventually harden. As you can figure below, the liquid wax and plastic have mixed, and on cooling, have formed what looks like a single solid substance. It is most likely that this is a very fine-grained mixture of solid wax and solid plastic, but it could also be some other substance that has formed from the combination of the two.

As depicted in figure above Partial melting of "pretend rock": (a) the original components of white candle wax, black plastic pipe, green beach glass, and aluminum wire, (b) after heating to 50 °C for 30 minutes only the wax has melted, (c) after heating to 120 °C for 60 minutes much of the plastic has melted and the two liquids have mixed, (d) the liquid has been separated from the solids and allowed to cool to make a "pretend rock" with a different overall composition.

In this example, we partially melted some pretend rock to create some pretend magma. We then separated the magma from the source and allowed it to cool to make a new pretend rock with a composition quite different from the original material (it lacks glass and aluminum).

Of course partial melting in the real world isn't exactly the same as in our pretend-rock example. The main differences are that rocks are much more complex than the four-component system we used, and the mineral components of most rocks have more similar melting temperatures, so two or more minerals are likely to melt at the same time to varying degrees. Another important difference is that when rocks melt, the process takes thousands to millions of years, not the 90 minutes it took in the pretend-rock example.

Contrary to what one might expect, and contrary to what we did to make our pretend rock, most partial melting of real rock does not involve heating the rock up. The two main mechanisms through which rocks melt are decompression melting and flux melting. Decompression melting takes place within Earth when a body of rock is held at approximately the same temperature but the pressure is reduced. This happens because the rock is being moved toward the surface, either at a mantle plume (a.k.a., hot spot), or in the upwelling part of a mantle convection cell. The mechanism of decompression melting is shown in figure. If a rock that is hot enough to be close to its melting point is moved toward the surface, the pressure is reduced, and the rock can pass to the liquid side of its melting curve. At this point, *partial* melting starts to take place. The process of flux melting is shown in figure. If a rock is close to its melting point and some water (a flux that promotes melting) is added to the rock, the melting temperature is reduced (solid line versus dotted line), and partial melting starts.

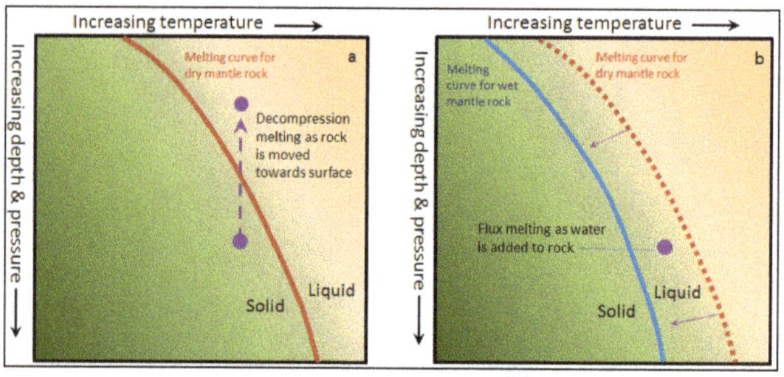

Figure: Mechanisms for (a) decompression melting (the rock is moved toward the surface) and (b) flux melting (water is added to the rock) and the melting curve is displaced.

The partial melting of rock happens in a wide range of situations, most of which are related to plate tectonics. The more important of these are shown in figure above. At both mantle plumes and in the upward parts of convection systems, rock is being moved toward the surface, the pressure is dropping, and at some point, the rock crosses to the liquid side of its melting curve. At subduction zones, water from the wet, subducting oceanic crust is transferred into the overlying hot mantle. This provides the flux needed to lower the melting temperature. In both of these cases, only partial melting takes place — typically only about 10% of the rock melts — and it is always the most silica-rich components of the rock that melt, creating a magma that is more silica-rich than the rock from which it is derived. (By analogy, the melt from our pretend rock is richer in wax and plastic than the "rock" from which it was derived.) The magma produced, being less dense than the surrounding rock, moves up through the mantle, and eventually into the crust.

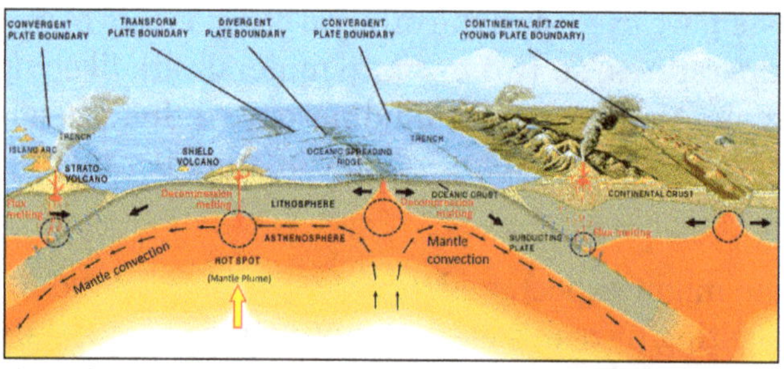

Figure: Common sites of magma formation in the upper mantle. The black circles are regions of partial melting. The blue arrows represent water being transferred from the subducting plates into the overlying mantle.

As it moves toward the surface, and especially when it moves from the mantle into the lower crust, the hot magma interacts with the surrounding rock. This typically leads to partial melting of the surrounding rock because most such magmas are hotter than the melting temperature of crustal rock. (In this case, melting is caused by an increase in temperature.) Again, the more silica-rich parts of the surrounding rock are preferentially melted, and this contributes to an increase in the silica content of the magma.

At very high temperatures (over 1300 °C), most magma is entirely liquid because there is too much energy for the atoms to bond together. As the temperature drops, usually because the magma is slowly moving upward, things start to change. Silicon and oxygen combine to form silica tetrahedra, and then, as cooling continues, the tetrahedra start to link together to make chains (polymerize). These silica chains have the important effect of making the magma more viscous.

Crystallisation of Magma

The minerals that make up igneous rocks crystallize at a range of different temperatures. This explains why a cooling magma can have some crystals within it and yet remain predominantly liquid. The sequence in which minerals crystallize from a magma is known as the Bowen reaction series.

Of the common silicate minerals, olivine normally crystallizes first, at between 1200° and 1300 °C. As the temperature drops, and assuming that some silica remains in the magma, the olivine crystals react (combine) with some of the silica in the magma to form pyroxene. As long as there is silica remaining and the rate of cooling is slow, this process continues down the discontinuous branch: olivine to pyroxene, pyroxene to amphibole, and (under the right conditions) amphibole to biotite.

At about the point where pyroxene begins to crystallize, plagioclase feldspar also begins to crystallize. At that temperature, the plagioclase is calcium-rich (anorthite). As the temperature drops, and providing that there is sodium left in the magma, the plagioclase that forms is a more sodium-rich variety.

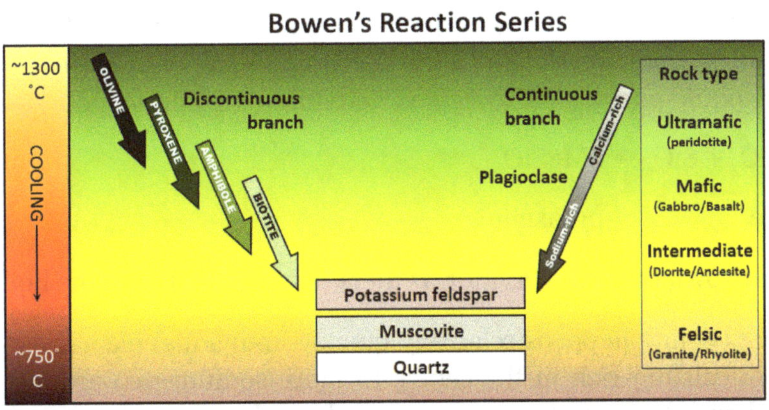

Figure: The Bowen reaction series describes the process of magma crystallization.

Bowen's Reaction Series

Norman Levi Bowen, born in Kingston Ontario, studied geology at Queen's University and then at MIT in Boston. In 1912, he joined the Carnegie Institution in Washington, D.C., where he carried out groundbreaking experimental research into the processes of cooling magmas. Working mostly with basaltic magmas, he determined the order of crystallization of minerals as the temperature drops. The method, in brief, was to melt the rock to a magma in a specially made kiln, allow it to cool slowly to a specific temperature (allowing some minerals to form), and then quench it (cool it quickly) so that no new minerals form (only glass). The results were studied under the microscope and by chemical analysis. This was done over and over, each time allowing the magma to cool to a lower temperature before quenching.

The Bowen reaction series is one of the results of his work, and even a century later, it is an important basis for our understanding of igneous rocks. The word *reaction* is critical. In the discontinuous branch, olivine is typically the first mineral to form (at just below 1300 °C). As the temperature continues to drop, olivine becomes unstable while pyroxene becomes stable. The early-forming olivine crystals *react* with silica in the remaining liquid magma and are converted into pyroxene, something like this:

$$\underset{\text{olivine}}{Mg_2SiO_4} + SiO_2 \rightarrow \underset{\text{pyroxene}}{2MgSiO_3}$$

This continues down the chain, as long as there is still silica left in the liquid.

In cases where cooling happens relatively quickly, individual plagioclase crystals can be zoned from calcium-rich in the centre to more sodium-rich around the outside. This occurs when calcium-rich early-forming plagioclase crystals become coated with progressively more sodium-rich plagioclase as the magma cools. The figure below shows a zoned plagioclase under a microscope.

Figure: A zoned plagioclase crystal. The central part is calcium-rich and the outside part is sodium-rich.

Finally, if the magma is quite silica-rich to begin with, there will still be some left at around 750° to 800 °C, and from this last magma, potassium feldspar, quartz, and maybe muscovite mica will form.

The composition of the original magma is critical to magma crystallization because it determines how far the reaction process can continue before all of the silica is used up. The compositions of typical mafic, intermediate, and felsic magmas are shown in Figure. Note that, unlike Figure, these compositions are expressed in terms of "oxides" (e.g., Al_2O_3 rather than just Al). There are two reasons for this: one is that in the early analytical procedures, the results were always expressed that way, and the other is that all of these elements combine readily with oxygen to form oxides.

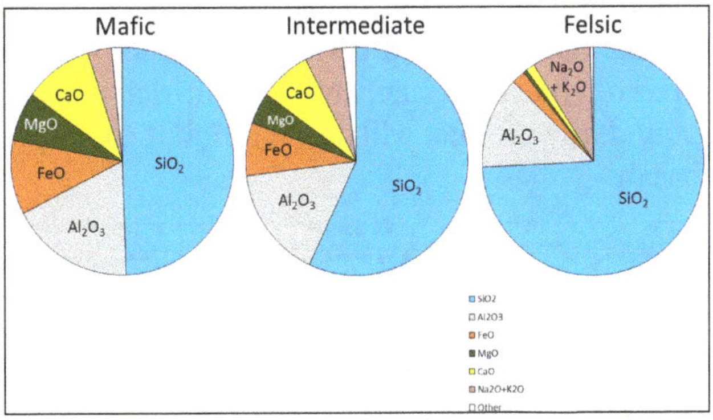

Figure: The chemical compositions of typical mafic, intermediate, and felsic magmas and the types of rocks that form from them.

Mafic magmas have 45% to 55% SiO_2, about 25% total of FeO and MgO plus CaO, and about 5% $Na_2O + K_2O$. Felsic magmas, on the other hand, have much more SiO_2 (65% to 75%) and $Na_2O + K_2O$ (around 10%) and much less FeO and MgO plus CaO (about 5%).

Example: Rock Types based on Magma Composition.

The proportions of the main chemical components of felsic, intermediate, and mafic magmas are listed in the table.

Oxide	Felsic Magma	Intermediate Magma	Mafic Magma
SiO_2	65-75%	55-65%	45-55%
Al_2O_3	12-16%	14-18%	14-18%
FeO	2-4%	4-8%	8-12%
CaO	1-4%	4-7%	7-11%
MgO	0-3%	2-6%	5-9%
Na_2O	2-6%	3-7%	1-3%
K_2O	3-5%	2-4%	0.5-3%

Chemical data for four rock samples are shown in the following table. Compare these with those in the table above to determine whether each of these samples is felsic, intermediate, or mafic.

SiO_2	Al_2O_3	FeO	CaO	MgO	Na_2O	K_2O	Type?
55%	17%	5%	6%	3%	4%	3%	
74%	14%	3%	3%	0.5%	5%	4%	
47%	14%	8%	10%	8%	1%	2%	
65%	14%	4%	5%	4%	3%	3%	

As a *mafic* magma starts to cool, some of the silica combines with iron and magnesium to make olivine. As it cools further, much of the remaining silica goes into calcium-rich plagioclase, and any silica left may be used to convert some of the olivine to pyroxene. Soon after that, all of the magma is used up and no further changes takes place. The minerals present will be olivine, pyroxene, and calcium-rich plagioclase. If the magma cools slowly underground, the product will be gabbro; if it cools quickly at the surface, the product will be basalt.

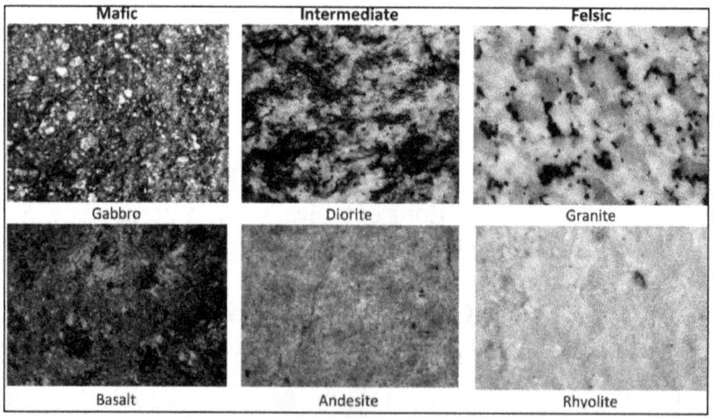

Figure: Examples of the igneous rocks that form from mafic, intermediate, and felsic magmas.

Felsic magmas tend to be cooler than mafic magmas when crystallization begins (because they don't have to be as hot to remain liquid), and so they may start out crystallizing pyroxene (not olivine) and plagioclase. As cooling continues, the various reactions on the discontinuous branch will proceed because silica is abundant, the plagioclase will become increasingly sodium-rich, and eventually potassium feldspar and quartz will form. Commonly even very felsic rocks will not have biotite or muscovite because they may not have enough aluminum or enough hydrogen to make the OH complexes that are necessary for mica minerals. Typical felsic rocks are granite and rhyolite.

The cooling behaviour of intermediate magmas lie somewhere between those of mafic and felsic magmas. Typical intermediate rocks are diorite and andesite.

A number of processes that take place within a magma chamber can affect the types of rocks produced in the end. If the magma has a low viscosity (i.e., it's runny) — which is likely if it is mafic — the crystals that form early, such as olivine, may slowly settle toward the bottom of the magma chamber. The means that the overall composition of the magma near the top of the magma chamber will become more felsic, as it is losing some iron- and magnesium-rich components. This process is known as fractional crystallization. The crystals that settle might either form an olivine-rich layer near the bottom of the magma chamber, or they might remelt because the lower part is likely to be hotter than the upper part.

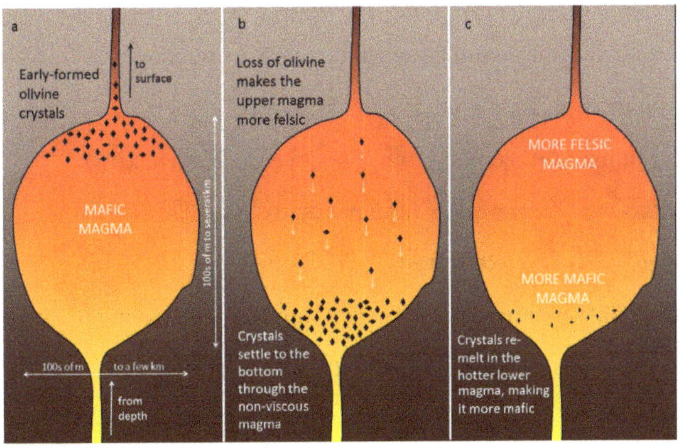

Figure: An example of crystal settling and the formation of a zoned magma chamber.

Figure: Porphyritic textures: volcanic porphyry (left – olivine crystals in Hawaiian basalt) and intrusive porphyry (right).

If crystal settling does not take place, because the magma is too viscous, then the process of cooling will continue as predicted by the Bowen reaction series. In some cases, however, partially cooled but still liquid magma, with crystals in it, will either move farther up into a cooler part of the crust, or all the way to the surface during a volcanic eruption. In either of these situations, the magma that has moved toward the surface is likely to cool much faster than it did within the magma chamber, and the rest of the rock will have a finer crystalline texture. An igneous rock with large crystals embedded in a matrix of finer crystals is indicative of a two-stage cooling process, and the texture is porphyritic.

References

- Igneous-rock, science: britannica.com, Retrieved 21 April, 2019
- Rhyolite, rocks: geology.com, Retrieved 20 January, 2019
- Granite-and-granodiorite: nps.gov, Retrieved 25 July, 2019
- Obsidian, rocks: geology.com, Retrieved 23 June, 2019
- Carbonatites: alexstrekeisen.it, Retrieved 1 May, 2019
- Diorite, rocks: geology.com, Retrieved 12 February, 2019
- Andesite, rocks: geology.com, Retrieved 14 July, 2019
- Diabase, science: britannica.com, Retrieved 2 June, 2019
- Tuff, rocks: geology.com, Retrieved 10 August, 2019
- Fire-opal, opal, gemstones: geology.com, Retrieved 11 March, 2019
- Gabbro: sandatlas.org, Retrieved 17 January, 2019
- Pumice, rocks: geology.com, Retrieved 27 April, 2019
- Magma-and-magma-formation, geology: opentextbc.ca, Retrieved 6 August, 2019
- Crystallization-of-magma, geology: opentextbc.ca, Retrieved 16 May, 2019

3

Sedimentary Rocks

The rocks which are formed by the accumulation of small particles and subsequent cementation of mineral or organic particles on the floor of bodies of water at the Earth's surface are known as sedimentary rocks. This chapter has been carefully written to provide an easy understanding of the varied types of sedimentary rocks such as breccia, conglomerate and siltstone.

Sedimentary rock is rock formed at or near the Earth's surface by the accumulation and lithification of sediment (detrital rock) or by the precipitation from solution at normal surface temperatures (chemical rock). Sedimentary rocks are the most common rocks exposed on the Earth's surface but are only a minor constituent of the entire crust, which is dominated by igneous and metamorphic rocks.

Sedimentary rocks are produced by the weathering of preexisting rocks and the subsequent transportation and deposition of the weathering products. Weathering refers to the various processes of physical disintegration and chemical decomposition that occur when rocks at the Earth's surface are exposed to the atmosphere (mainly in the form of rainfall) and the hydrosphere. These processes produce soil, unconsolidated rock detritus, and components dissolved in groundwater and runoff. Erosion is the process by which weathering products are transported away from the weathering site, either as solid material or as dissolved components, eventually to be deposited as sediment. Any unconsolidated deposit of solid weathered material constitutes sediment. It can form as the result of deposition of grains from moving bodies of water or wind, from the melting of glacial ice, and from the downslope slumping (sliding) of rock and soil masses in response to gravity, as well as by precipitation of the dissolved products of weathering under the conditions of low temperature and pressure that prevail at or near the surface of the Earth.

Sedimentary rocks are the lithified equivalents of sediments. They typically are produced by cementing, compacting, and otherwise solidifying preexisting unconsolidated sediments. Some varieties of sedimentary rock, however, are precipitated directly into their solid sedimentary form and exhibit no intervening existence as sediment. Organic reefs and bedded evaporites are examples of such rocks. Because the processes of physical (mechanical) weathering and chemical weathering are significantly different, they generate markedly distinct products and two fundamentally different kinds of sediment and sedimentary rock:

1. Terrigenous clastic sedimentary rocks and,
2. Allochemical and orthochemical sedimentary rocks.

Clastic terrigenous sedimentary rocks consist of rock and mineral grains, or clasts, of varying size, ranging from clay-, silt-, and sand- up to pebble-, cobble-, and boulder-size materials. These clasts are transported by gravity, mudflows, running water, glaciers, and wind and eventually are deposited in various settings (e.g., in desert dunes, on alluvial fans, across continental shelves, and in river deltas). Because the agents of transportation commonly sort out discrete particles by clast size, terrigenous clastic sedimentary rocks are further subdivided on the basis of average clast diameter. Coarse pebbles, cobbles, and boulder-size gravels lithify to form conglomerate and breccia; sand becomes sandstone; and silt and clay form siltstone, claystone, mudrock, and shale.

Chemical sedimentary rocks form by chemical and organic reprecipitation of the dissolved products of chemical weathering that are removed from the weathering site. Allochemical sedimentary rocks, such as many limestones and cherts, consist of solid precipitated nondetrital fragments (allochems) that undergo a brief history of transport and abrasion prior to deposition as nonterrigenous clasts. Examples are calcareous or siliceous shell fragments and oöids, which are concentrically, layered spherical grains of calcium carbonate. Orthochemical sedimentary rocks, on the other hand, consist of dissolved constituents that are directly precipitated as solid sedimentary rock and thus do not undergo transportation. Orthochemical sedimentary rocks include some limestones, bedded evaporite deposits of halite, gypsum, and anhydrite, and banded iron formations.

Sediments and sedimentary rocks are confined to the Earth's crust, which is the thin, light outer solid skin of the Earth ranging in thickness from 40–100 kilometres (25 to 62 miles) in the continental blocks to 4–10 kilometres in the ocean basins. Igneous and metamorphic rocks constitute the bulk of the crust. The total volume of sediment and sedimentary rocks can be either directly measured using exposed rock sequences, drill-hole data, and seismic profiles or indirectly estimated by comparing the chemistry of major sedimentary rock types to the overall chemistry of the crust from which they are weathered. Both methods indicate that the Earth's sediment-sedimentary rock shell forms only about 5 percent by volume of the terrestrial crust, which in turn accounts for less than 1 percent of the Earth's total volume. On the other hand, the area of outcrop and exposure of sediment and sedimentary rock comprises 75 percent of the land surface and well over 90 percent of the ocean basins and continental margins. In other words, 80–90 percent of the surface area of the Earth is mantled with sediment or sedimentary rocks rather than with igneous or metamorphic varieties. The sediment-sedimentary rock shell forms only a thin superficial layer. The mean shell thickness in continental areas is 1.8 kilometres; the sediment shell in the ocean basins is roughly 0.3 kilometre. Rearranging this shell as a globally encircling layer (and depending on the raw estimates incorporated into the model), the shell thickness would be roughly 1–3 kilometres.

Despite the relatively insignificant volume of the sedimentary rock shell, not only are most rocks exposed at the terrestrial surface of the sedimentary variety, but many of the significant events in Earth history are most accurately dated and documented by analyzing and interpreting the sedimentary rock record instead of the more voluminous igneous and metamorphic rock record. When properly understood and interpreted, sedimentary rocks provide information on ancient geography, termed paleogeography. A map of the distribution of sediments that formed in shallow oceans along alluvial fans bordering rising mountains or in deep, subsiding ocean trenches will indicate past relationships between seas and landmasses. An accurate interpretion of paleogeography and depositional settings allows conclusions to be made about the evolution of mountain systems, continental blocks, and ocean basins, as well as about the origin and evolution of the atmosphere and hydrosphere. Sedimentary rocks contain the fossil record of ancient life-forms that enables the documentation of the evolutionary advancement from simple to complex organisms in the plant and animal kingdoms. Also, the study of the various folds or bends and breaks or faults in the strata of sedimentary rocks permits the structural geology or history of deformation to be ascertained.

Finally, it is appropriate to underscore the economic importance of sedimentary rocks. For example, they contain essentially the world's entire store of oil and natural gas, coal, phosphates, salt deposits, groundwater, and other natural resources.

Several subdisciplines of geology deal specifically with the analysis, interpretation, and origin of sediments and sedimentary rocks. Sedimentary petrology is the study of their occurrence, composition, texture, and other overall characteristics, while sedimentology emphasizes the processes by which sediments are transported and deposited. Sedimentary petrography involves the classification and study of sedimentary rocks using the petrographic microscope. Stratigraphy covers all aspects of sedimentary rocks, particularly from the perspective of their age and regional relationships as well as the correlation of sedimentary rocks in one region with sedimentary rock sequences elsewhere.

Classification Systems

In general, geologists have attempted to classify sedimentary rocks on a natural basis, but some schemes have genetic implications (i.e. knowledge of origin of a particular rock type is assumed), and many classifications reflect the philosophy, training, and experience of those who propound them. No scheme has found universal acceptance, and discussion here will centre on some proposals.

The book Rocks and Rock Minerals by Louis V. Pirsson was first published in 1908, and it has enjoyed various revisions. Sedimentary rocks are classified there rather simplistically according to physical characteristics and composition into detrital and nondetrital rocks.

Numerous other attempts have been made to classify sedimentary rocks. The most significant advance occurred in 1948 with the publication in the Journal of Geology of three definitive articles by the American geologists Francis J. Pettijohn, Robert R.

Shrock, and Paul D. Krynine. Their classifications provide the basis for all modern discussion of the subject. The nomenclature associated with several schemes of classifying clastic and nonclastic rocks will be discussed in the following sections, but a rough division of sedimentary rocks based on chemical composition is shown in figure.

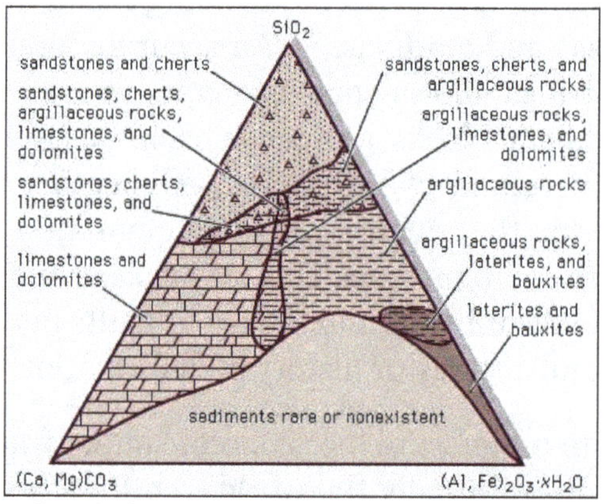

Chemical composition of sedimentary rocks.

For the purposes of the present discussion, three major categories of sedimentary rocks are recognized:

1. Terrigenous clastic sedimentary rocks,
2. Carbonates (limestone and dolomite),
3. Noncarbonate chemical sedimentary rocks.

Terrigenous clastic sedimentary rocks are composed of the detrital fragments of preexisting rocks and minerals and are conventionally considered to be equivalent to clastic sedimentary rocks in general. Because most of the clasts are rich in silica, they are also referred to as siliciclastic sedimentary rocks. Siliciclastics are further subdivided on the basis of clast diameter into conglomerate and breccia, sandstone, siltstone, and finer-than-silt-sized mudrock (shale, claystone, and mudstone). The carbonates, limestones and dolomites, consist of the minerals aragonite, calcite, and dolomite. They are chemical sedimentary rocks in the sense that they possess at least in part a crystalline, interlocking mosaic of precipitated carbonate mineral grains. However, because individual grains such as fossil shell fragments exist for some period of time as sedimentary clasts, similar to transported quartz or feldspar clasts, most carbonates bear some textural affinities to the terrigenous clastic sedimentary rocks. The noncarbonate chemical sedimentary rocks include several rock types that are uncommon in the sedimentary rock record but remain important either from an economic point of view or because their deposition requires unusual settings. Specific varieties discussed below include siliceous rocks (cherts), phosphate rocks (phosphorites), evaporites, iron-rich sedimentary rocks (iron formations and ironstones), and organic-rich (carbonaceous) deposits in sedimentary rocks (coal, oil shale, and petroleum).

Despite the diversity of sedimentary rocks, direct measurement of the relative abundance of the specific types based on the study of exposed sequences suggests that only three varieties account for the bulk of all sedimentary rocks: mudrock, 47 percent; sandstone, 31 percent; and carbonate, 22 percent. Another method, which involves comparing the chemical composition of major sedimentary rock types with the chemistry of the Earth's continental crust, yields somewhat different numbers: mudrock, 79 percent; sandstone, 13 percent; and carbonate, 8 percent. Most sedimentary petrologists concede that the sedimentary rock record preserved and exposed within the continental blocks is selectively biased in favour of shallow-water carbonates and sandstones. Mudrocks are preferentially transported to the ocean basins. Consequently, indirect estimates based on chemical arguments are probably more accurate.

Terrigenous Clastic Rocks

A prominent physical feature of terrigenous clastic rocks is texture—that is, the size, shape, and arrangement of the constituent grains. These rocks have a fragmental texture: discrete grains are in tangential contact with one another. Terrigenous clastic sedimentary rocks are further subdivided on the basis of the mean grain diameter that characterizes most fragments, using the generally accepted size limits. Granules, pebbles, cobbles, boulders, and blocks constitutethe coarse clastic sediments; sand-size (arenaceous) clasts are considered medium clastic sediments; and fine clastics sediments consists of silt- and clay-size materials.

The simplest way of classifying coarse clastic sedimentary rocks is to name the rock and include a brief description of its particular characteristics. Conglomerates and breccias differ from one another only in clast angularity. The former consist of abraded, somewhat rounded, coarse clasts, whereas the latter contain angular, coarse clasts. Thus, a pebble conglomerate is a coarse clastic sedimentary rock whose discrete particles are rounded and range from 4 to 64 millimetres (0.2 to 2.5 inches) in diameter. A more precise description reveals the rock types of the mineral fragments that compose the conglomerate—for example, a granite-gneiss pebble conglomerate.

Sandstones have long intrigued geologists because they are well exposed, are abundant in the geologic record, and provide an enormous amount of information about depositional setting and origin. Many classification schemes have been developed for sandstones, only the most popular of which are reviewed below. Most schemes emphasize the relative abundance of sand-size quartz, feldspar, and rock fragment components, as well as the nature of the material housed between this sand-size "framework" fractions.

Fine clastics are commonly, but rather simplistically, referred to as mudrocks. Mudrocks actually can include any clastic sedimentary rock in which the bulk of the clasts have diameters finer than 1/16 millimetre. Varieties include siltstone (average grain size between 1/16and 1/256 millimetre) and claystone (discrete particles are mostly finer than 1/256 millimetre). Mudis a mixture of silt- and clay-size material, and mudrock is its indurated product. Shale is any fine clastic sedimentary rock that exhibits fissility, which is the ability to break into thin slabs along narrowly spaced planes parallel to the layers of stratification. Despite the great abundance of the fine clastics, disagreement exists as to what classification schemes are most useful for them, and an understanding of their origin is hindered by analytical complexities.

Carbonate Rocks: Limestones and Dolomites

Limestones and dolostones (dolomites) make up the bulk of the nonterrigenous sedimentary rocks. Limestones are for the most part primary carbonate rocks. They consist of 50 percent or more calcite and aragonite (both $CaCO_3$). Dolomites are mainly produced by the secondary alteration or replacement of limestones; i.e., the mineral dolomite [$CaMg(CO_3)_2$] replaces the calcite and aragonite minerals in limestones during diagenesis. A number of different classification schemes have been proposed for carbonates, and the many categories of limestones and dolomites in the geologic record represent a large variety of depositional settings.

Noncarbonate Chemical Sedimentary Rocks

Noncarbonate chemical sedimentary rocks differ in many respects from carbonate sedimentary rocks and terrigenous clastic sedimentary rocks, and there is no single classification that has been universally accepted. This is a reflection of the great variation in mineral composition, texture, and other properties of these rock types. Such rocks as ironstones and banded iron formations (limonite, goethite, hematite, siderite, and chamosite), phosphorites, evaporites (rock salt, gypsum, and other salts), siliceous rocks (cherts), and organic-rich (carbonaceous) deposits of oil, natural gas, and coal in sedimentary rocks occur in much less abundance than carbonates and siliciclastic sedimentary rocks, although they may form thick and widespread deposits.

Classification schemes that incorporate all types of noncarbonate chemical sedimentary rocks do not exist because no triangular or tetrahedral scheme can accommodate all of them. Several of the major types are shown in the tetrahedron in figure.

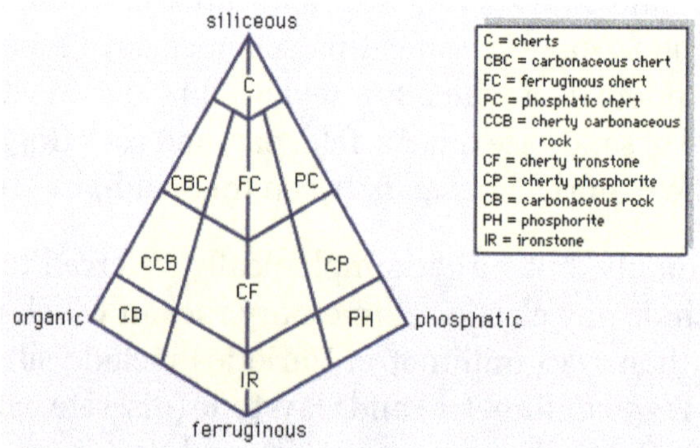

Common noncarbonated: Nonclastic sedimentary rocks.

Properties of Sedimentary Rocks

Texture

Texture refers to the physical makeup of rock—namely, the size, shape, and arrangement (packing and orientation) of the discrete grains or particles of a sedimentary rock. Two main natural textural groupings exist for sedimentary rocks: clastic (or fragmental) and nonclastic (essentially crystalline). Noncarbonate chemical sedimentary rocks in large part exhibit crystalline texture, with individual mineral grains forming an interlocking arrangement. Depositional setting is an insignificant factor in both determining crystal size and altering crystalline texture. The size of crystals is controlled to a greater degree by the rate of precipitation, and their texture is modified by postdepositional recrystallization (reflecting the diagenetic environment). As a result, little attention is paid to crystalline textures other than providing a simple description of it (for example, coarsely crystalline versus finely crystalline). Also, even though carbonate rocks commonly include allochems that behave as clasts, they too are commonly diagenetically altered. Consequently, only cursory efforts are made to texturally characterize limestones and dolomites. Therefore, the following discussion deals in detail only with the textural techniques applied to terrigenous (siliciclastic) sedimentary rocks.

Grain size

Particle size is an important textural parameter of clastic rocks because it supplies information on the conditions of transportation, sorting, and deposition of the sediment and provides some clues to the history of events that occurred at the depositional site prior to final induration. Determining the sizes of the discrete particles that constitute a sedimentary rock can be difficult, particularly if the rock is firmly indurated (cemented, compacted, and lithified). Various methods of measuring grain-size distribution have been devised; likewise several different grade-size schemes exist.

The size of particulate materials that make up sediments and sedimentary rocks are measured by weighing the proportions that accumulate in a series of wire mesh screen sieves, by visually counting grains with a petrographic microscope, or by determining the rate at which particles of varying diameter accumulate in a water-filled glass cylinder (known as a settling tube).

After the grain-size distribution for a given sediment or sedimentary rock has been determined by sieving, microscopic analysis, or use of a settling tube, it can be characterized using standard statistical measures in either of two ways:

1. Visual inspection of various types of graphs that plot overall percent abundance versus grain-size diameter (e.g., histograms or bar diagrams, size frequency and cumulative size frequency curves, and probability curves that compare the actual grain-size distribution to a normal straight-line Gaussian distribution) or

2. Arithmetic calculations made using diameter values in either millimetres or phi units that are read off the graphic plots and inserted into standard formulas.

For siliciclastic sedimentary rocks, the following standard statistical measures are conventionally described for grain-size distributions:

1. Mode, the most frequently occurring particle size or size class,
2. Median, the midpoint size of any grain-size distribution,
3. Mean, an estimate of the arithmetic average particle size,
4. Sorting or standard deviation, a measure of the range, scatter, or variation in grain size,
5. Skewness, the degree of symmetry or asymmetry of the grain-size distribution, which is in turn a function of the coincidence or noncoincidence of mean, median, and mode, and
6. Kurtosis (peakedness) of a grain-size distribution, which compares sorting in the central portion of the population with that in the tails.

Analysis of grain-size distribution is conducted with the disputed assumption that particular transporting agents and depositional settings (e.g., river delta deposits versus shallow marine longshore-bar sands) impose a distinctive textural "fingerprint" on the sediments they produce. Despite continuing efforts, the success of the various graphic and arithmetic approaches in characterizing grain-size distributions is debatable, as is their reliability in pinpointing ancient depositional settings. The grain-size distribution of sediments in many settings commonly appears to be inherited or to exhibit as much variation within a single environment as between different ones.

Particle Shape

Three different but related properties determine particle shape: form, roundness, and surface texture. Particle form is the overall shape of particles, typically defined in terms of the relative lengths of the longest, shortest, and intermediate axes. Particles can be spherical, prismatic, or bladelike. Roundness or angularity is a measure of the smoothness of particles. Surface texture refers to the presence or absence of small, variously shaped markings (pits, polish, scratches) that may occur on grain surfaces.

Each of these attributes of particle shape is traditionally measured in a standard fashion for the purpose of identifying the transporting agent and the depositional environment. Form is determined either by painstakingly measuring individual particles in three dimensions or by Fourier shape analysis, which uses harmonics analysis and computer digitizing to provide a precise description of particles in two dimensions. Form alone has limited usefulness in inferring depositional setting but more accurately reflects the mineralogy of the grains involved. Roundness is characterized by visually comparing grains to standard silhouette profiles. It is largely the result of abrasion history, which is controlled by the depositional agent and environment. For example, windblown and surf zone sands are well-rounded, while glacial sands and turbidity current deposits are angular. Particle roundness or angularity also reflects mineralogy (soft minerals are abraded more readily than hard minerals), clast size (coarse particles become rounded more rapidly than do fine ones), and transport distance (sands become more abraded and hence rounder as the distance traveled increases). Particle surfaces can be visually examined for pitting, markings, and polish through the use of a microscope or hand lens, or in some cases, a scanning electron microscope (SEM). Certain surface textures have been genetically linked to specific depositional agents; for example, classic V-shaped percussion marks identify quartz grains of the beach and nearshore zones.

Fabric

The fabric of a sedimentary rock controls the rock's porosity and permeability and therefore its ability to hold and transmit fluids such as oil and water. The orientation, or lack thereof, of the crystals or grains that make up a sedimentary rock constitutes one aspect of fabric. Genetically, there are two principal varieties of oriented fabrics: primary (or depositional) and secondary (or deformational). Primary fabrics are produced while the sediment is accumulating. For example, river currents and some submarine gravity flows generate sediments whose flaky and prismatic constituent particles have long or short axes parallel with one another to produce an oriented fabric. Secondary fabrics result from a rotation of the constituent elements under stress or from the growth of new elements during diagenesis. Fabrics in coarse clastic sedimentary rocks like conglomerates and sandstones can be determined by measuring and plotting dimensional directions, such as the long axes of pebbles or sand grains. In mudrocks, fabrics can be ascertained by studying the platelike arrangement of mica and clay minerals.

In addition to orientation, a factor known as packing contributes to a rock's fabric. Packing refers to the distribution of grains and intergranular spaces (either empty or filled with cement or fine-grained matrix) in a sedimentary rock. It is controlled by grain size and shape and by the degree of compaction of a sedimentary rock; in turn it determines the rock's bulk density. A description of packing is generally based on the analysis of thin sections of a sedimentary rock using a petrographic microscope. Particular attention is paid to the number of grain-to-grain contacts (packing proximity) and to comparisons between the sums of the lengths of grains to the total length of a traverse across a thin section (packing density).

Mineralogical and Geochemical Composition

Minerals that make up sedimentary rocks are of two principal types—namely, detrital and authigenic. Detrital minerals, such as grains of quartz and feldspar, survive weathering and are transported to the depositional site as clasts. Authigenic minerals, like calcite, halite, and gypsum, form in situ within the depositional site in response to geochemical processes. The chemical compounds that constitute them ultimately are generated by chemical weathering and are transported from the weathering site to the point of precipitation primarily in solution. Clay minerals are abundant in sedimentary rocks, particularly mudrocks, and some are detrital. They may have been produced at the weathering site by the partial decomposition of minerals like feldspar. They are transported as clasts, however, and thus can be regarded simply as fine- to very fine-textured detrital particles. Other clay minerals form authigenically at the site of deposition. Some of the important clay minerals are kaolinite, halloysite, montmorillonite, illite, vermiculite, and chlorite.

The mean chemical composition of the major varieties of sedimentary rocks exhibits wide variation. Significant contrasts in overall composition among sandstones, carbonates, and mudrocks reflect fundamental differences not only in the mechanisms by which detrital minerals of different sizes are transported and deposited but also in the chemical conditions that permit precipitation of various authigenic minerals.

Diagenesis includes all physicochemical, biochemical, and physical processes (short of metamorphism) that modify sediments in the time between their deposition and their analysis. Lithification, the process by which sediment is converted into solid sedimentary rock, is one result of diagenesis. Many diagenetic processes such as cementation, recrystallization, and dolomitization are essentially geochemical processes; others like compaction are fundamentally physical processes. All diagenetic changes occur at the low temperatures and pressures characteristic of surface and near-surface environments. These changes can take place almost immediately after sediment formation, or they can occur hundreds or even millions of years later.

Sedimentary Structures

Sedimentary structures are the larger, generally three-dimensional physical features of sedimentary rocks; they are best seen in outcrop or in large hand specimens rather than through a microscope. Sedimentary structures include features like bedding, ripple marks, fossil tracks and trails, and mud cracks. They conventionally are subdivided into categories based on mode of genesis. Structures that are produced at the same time as the sedimentary rock in which they occur are called primary sedimentary structures. Examples include bedding or stratification, graded bedding, and cross-bedding. Sedimentary structures that are produced shortly after deposition and as a result of compaction and desiccation are called penecontemporaneous sedimentary structures. Examples include mud cracks and load casts. Still other sedimentary structures like concretions, vein fillings, and stylolites form well after deposition and penecontemporaneous modification; these are known as secondary structures. Finally, others like stromatolites and organic burrows and tracks, though they may in fact be primary, penecontemporaneous, or even secondary, may be grouped as a fourth category—organic sedimentary structures.

Considerable attention is paid to the sedimentary structures exhibited by any sedimentary rock. Primary sedimentary structures are particularly useful because their abundance and size suggest the probable transporting and depositional agents. Certain varieties of primary sedimentary structures like cross-bedding and ripple marks display orientations that are consistently related to the direction of current movement. Such structures are referred to as directional sedimentary structures because they can be used to infer the ancient paleocurrent pattern or dispersal system by which a sedimentary rock unit was deposited. Other sedimentary structures are stratigraphic "top and bottom" indicators. For example, the progressive upward decrease in clastic grain size diameters, known as graded bedding, would allow a geologist to determine which way is stratigraphically "up"—i.e., toward the younger beds in a dipping sedimentary bed. The suite (repeated sequence) of sedimentary structures in any single stratigraphic unit is another attribute by which that unit may be physically differentiated from others in the region.

External Stratification

Stratification (or bedding) is expressed by rock layers (units) of a general tabular or lenticular form that differ in rock type or other characteristics from the material with which they are interstratified (sometimes stated as interbedded, or interlayered). These beds, or strata, are of varying thickness and areal extent. The term stratum identifies a single bed, or unit, normally greater than one centimetre in thickness and visibly separable from superjacent (overlying) and subjacent (underlying) beds. "Strata" refers to two or more beds, and the term lamina is sometimes applied to a unit less than one centimetre in thickness. Thus, lamination consists of thin units in bedded, or layered, sequence in a natural rock succession, whereas stratification consists of bedded layers, or strata, in a geologic sequence of interleaved sedimentary rocks.

For most stratified sedimentary rocks, the arrangement of layers is one of unequal thickness, ranging from very thin laminae to discrete beds that measure a few to many metres in thickness. The terms thick and thin as applied to bedding, or stratification, are relative, reflecting the training of a particular geologist as well as experience with a specific stratigraphic section or sections.

Bedding Types and Bedding-plane Features

It is common to discover a rhythmic pattern in a pile of stratified sedimentary rocks represented by a repetitive sequence of rock types. In most instances of such cyclic sedimentation, the bedding, or stratification, is horizontal or essentially so; that is, the transporting, sorting, and depositing agents of wind, running water, and lake and ocean currents and waves accumulated the laminae and strata in a flat-lying or horizontal arrangement. They are termed well-bedded, a type of primary stratification.

Primary stratification in sediments and sedimentary rocks can be cross-bedded (cross-stratified), graded, and imbricate and can also display climbing laminae, ripples, and beds.

Graded bedding simply identifies strata that grade upward from coarse-textured clastic sediment at their base to finer-textured materials at the top. The stratification may be sharply marked so that one layer is set off visibly from those above and beneath it. More commonly, however, the layers are blended. This variety of bedding results from a check in the velocity of the transporting agent, and thus coarse-textured sediment (gravel, for example) is deposited first, followed upward by pebbles, granules, sand, silt, and clay. It is commonly associated with submarine density currents.

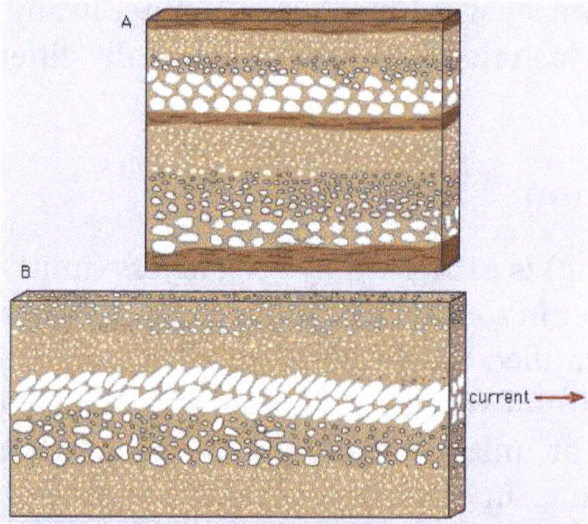

(A) Graded bedding. (B) Imbricate bedding.

Imbricate bedding is a shingle structure in a deposit of flattened or disk-shaped pebbles or cobbles. That is to say, elongated and commonly flattened pebbles and cobbles in gravelly sediment are deposited so that they overlap one another like roofing shingles. Imbricate bedding forms where high-velocity currents move over a streambed or where strong currents and waves break over a gradually sloping beach, thereby forming beach shingle.

Growth structures in sedimentary rocks are in situ features that accumulate largely as the result of organic buildups within otherwise horizontal or nearly flat-lying strata. Reefs and stromatolites are two common varieties of such growth structures.

Bedding-plane Features

Upper surfaces of beds commonly display primary sedimentary features that are classified as bedding-plane structures. A three-dimensional view may be obtained if some of these can be seen from the side as well as from the top of a pile of strata. They include such features as ripples (ripple marks), climbing ripples, rills, pits, mud cracks, trails and tracks, salt and ice casts and molds, and others. Bedding-plane markings and irregularities can be allocated to one of three classes.

1. Those on the base of a bed (load and current structures and organic markings),
2. Those within a bed (parting lineation), and
3. Those on top of a bed (ripple marks, pits, impressions, mud cracks, tracks and trails of organisms, and others).

Deformation Structures

In addition to sedimentary structures that are normally associated with bedding planes, there are other such structures that result from deformation during or shortly after sedimentation but before induration of the sediment into rock. These are non-tectonic features—i.e., they are not bends and folds brought about by metamorphism or other such causes. Deformation structures can be grouped into several classes, as follows:

1. Founder and load structures,
2. Convoluted structures,
3. Slump structures,
4. Injection structures, such as sandstone dikes or sills, and
5. Organic structures.

Structures found on the bottom of a bed are called sole markings, because they formed on the "sole" of the bed. Sole marks are commonly formed on sandstone and limestone beds that rest upon shale beds. They are termed casts, because they are fillings of depressions that formed on the surface of the underlying mud. They originate (1) by unequal loading upon the soft and plastic wet mud, (2) by the action of currents across the upper mud surface, or (3) by the activities of organisms on this surface. Load casts form as the result of downsinking of sandstone or limestone into the mud beneath. Current marks can form by the action of water currents on upper surfaces of the beds or by "tools" (such as wood and fossils) that are transported by currents over soft sediment.

Sedimentary Environments

The sedimentary environment is the specific depositional setting of a particular sedimentary rock and is unique in terms of physical, chemical, and biological characteristics. The physical features of a sedimentary environment include water depth and the velocity and persistence of currents. Chemical characteristics of an environment include the salinity (proportion of dissolved salts), acidity or basicity (pH), oxidation potential (Eh), pressure, and temperature. The biological characteristics are mainly the assemblage of fauna and flora that populate the setting. These conditions, combined with the nature of the transporting agent and the source area, largely determine the properties of the sediments deposited within the environment. A number of ways of classifying depositional environments exist, but most modern schemes employ a geomorphologic approach. That is to say, an environment is defined in terms of a distinct geomorphic unit or landform, modern examples of which are readily visible for comparative purposes—e.g., a river delta, an alluvial fan, a submarine fan, or the abyssal floor of an ocean basin.

Individual environments are further grouped into:

1. Marine environments, which include the nearshore, shallow littoral zone and the offshore, deep littoral zone, as well as deepwater realms,
2. Mixed marine and nonmarine settings such as the beach and supratidal zones, and
3. Nonmarine settings like lacustrine and various alluvial settings. Each environment is associated with a set of criteria that constitutes its distinguishing features.

Secular Trends in the Sedimentary Rock Record

Reexamination of the sedimentary rock record preserved within the continental blocks suggests systematic changes through time in the relative proportions of the major sedimentary rock types deposited, as summarized in Figure. These changes can be linked to the evolution of the atmosphere and hydrosphere and to the changing global tectonic setting. Carbonates and quartz sands, for example, require long-term source area stability as well as the existence of broad, shallow-water epeiric seas that mantle continental blocks. Marine transgressions and regressions across broad stable continental cratons occurred only in Proterozoic and Phanerozoic time; Archean continental blocks were smaller and tectonically unstable, and most likely less granitic than those of today. Consequently, the early Precambrian sedimentary rock record consists largely of volcanogenic sediments, wackes, and arkoses physically disintegrated from small, high-relief island arcs (the Archean greenstone belts of the various Precambrian shields) and microcontinental fragments. The fact that iron formations are restricted to rocks of Archean and Proterozoic time supports the conclusion that atmospheric oxygen levels in earlier stages of Earth history were lower, promoting the dissolution, transport, and precipitation of iron by chemical or biochemical means. The total lack of evaporites from the Archean record and their subsequent steady buildup probably reflects a number of factors. Such deposits can be easily destroyed by metamorphism, and presumably, given enough time, they will have been completely erased. Also, evaporite formation requires seawater with elevated salinity, a condition that is established with time. Finally, significant volumes of bedded evaporites can occur only once broad, stable continents have evolved, because the requisite restricted evaporite basins develop exclusively adjacent to cratons.

Sedimentary Environments

A sedimentary, or depositional, environment is an area on Earth's surface, such as a lake or stream, where large volumes of sediment accumulate. All environments of deposition belong to one of three settings: terrestrial, coastal (or marginal marine), and marine. Subenvironments, each with their own characteristic environmental factors and sedimentary deposits, make up a sedimentary environment. For example, streams consist of channel, sand bar, levee, and floodplain subenvironments, among others.

Sedimentary environments display great complexity and almost infinite variety. Variations in environmental factors such as climate, latitude, surface topography, subsurface geology, and sediment supply help determine the characteristics of a particular sedimentary environment, and the resulting sedimentary deposits. This entry deals only with typical examples of common environments, with greatly simplified descriptions.

Terrestrial Environments

Water, wind, and ice erode, transport, and deposit terrigenous sediments on land. Geologists recognize five common terrestrial sedimentary environments: stream, lake, desert, glacial, and volcanic.

Streams are the most widespread terrestrial sedimentary environment. Because they dominate landscapes in both humid and arid climates, stream valleys are the most common landform on Earth. Streams naturally meander and coarse-grained sediments accumulate along the inside of meanders where water velocity decreases, forming sand and gravel bars. When flood-water overflows a stream's banks, fine-grained sediment accumulates on the land surface, or floodplain, adjacent to the channel. Coarser sediment collects on the channel banks during floods, forming a narrow deposit called a levee. Sorting, rounding, and sediment load generally increase downstream.

Where a stream rapidly changes from a high to low slope on land, for example at the base of a mountain, gravel, sand, silt, and clay form a sediment pile called an alluvial fan. Where a stream flows into standing water its sediments produce a deposit called a delta. Deltas are usually finer grained than alluvial fans. In both alluvial fans and deltas, grain size rapidly decreases downslope.

Most lakes form from water contributed by one or more streams as well as precipitation directly into the lake. As it arrives at a lake, stream velocity drops very rapidly, depositing the coarsest sediment at the lake-shore and forming a delta. Farther from shore, as the water continues to lose velocity, finer and finer grained sediment falls to the lake bottom. Only in the deepest part of the lake is water movement slow enough to permit the finest grained sediment to accumulate. This produces thin layers of clay. Hence, grain size generally decreases from the lakeshore to its center.

Deserts develop where rainfall is too sparse to support abundant plants. Contrary to popular belief, most deserts are not vast seas of sand. Instead, they consist mostly of a mixture of gravel and sand. However, the sand may be eroded away, or deflated, by the wind leaving behind a layer of gravel called a desert pavement, or reg. The deflated sand is later heaped into piles downwind, producing dunes. Despite the prevalence of regs and dunes in deserts, water is nonetheless the most important agent of erosion. Alluvial fans are common at the base of mountains. Dry lake beds, or playas, and salt deposits, or sabkhas, resulting from lake evaporation, commonly occupy the adjacent valley floor.

Where snowfall exceeds snowmelt and snow persists from year to year, ice accumulation can eventually form a glacier. Alpine glaciers occur throughout the world on mountains at high elevations. Modern continental glaciers now cover Antarctica and Greenland.

From around two million years ago to about ten thousand years ago—the Pleistocene epoch, or Ice Age—glaciers deposited sediments over large areas at mid- to high-latitudes. These glacial ice deposits, called till, are characterized by a wide range of sediment sizes (geologists refer to sediment comprised of a wide range of particle sizes as being poorly sorted). They generally are thick, widespread sheets or narrow, sinuous ridges. Ice meltwater forms thick, well-sorted, and widespread layers of sediment called stratified drift.

Though volcanism involves igneous processes, many terrestrial volcanic deposits are sedimentary in origin. These volcaniclastic, or pyroclastic, sediments form when ash, cinders, and larger volcanic materials fall to the ground during eruptions. Running water often modifies volcaniclastic sediments after deposition. They also may move downhill as a mudflow, or lahar, when saturated with water. Generally, volcani-clastic sediments form thin lobe-shaped deposits and widespread sheets, which thicken toward the volcanic source.

Coastal Environments

Where the land meets the sea, interplay between terrestrial and marine processes causes sedimentary environments to be complicated. In areas where wave energy is low and the tidal range (the difference between high tide and low tide) is also low, terrestrial processes usually dominate. For example, sediments flowing into the sea from a river will form a well-developed triangular shaped deposit known as a delta. If wave energy is high and tidal range low, the river's sediments will be reworked into a beach or barrier island. However, if tidal range is high, tidal currents flood the river mouth daily, forming a drowned river mouth, or estuary, with scattered sand bars.

In coastal areas far from rivers, the nature of the coast changes rapidly. The balance between tidal and wave processes influences coastal character. The higher the tidal range, the more important tidal processes become. In wave-dominated areas, currents flowing parallel to the shoreline move sand along the coast, producing barrier islands and beaches for long stretches. If a barrier island protects the coast, channels, or tidal inlets, pass between the islands and allow tidal currents to flow from the open ocean into the bay behind the island. Landward from the bay, a tidal marsh will occur. When high tide approaches and tidal currents flow landward, the marsh will be flooded. As the water level drops toward low tide, tidal currents flow seaward, exposing the marsh to the elements. If no barrier island is present, coasts are simple with only a few river mouths and coastal marshes to break the monotony of long stretches of beach.

Where tidal range is high, strong tidal currents dominate coastal processes. Tidal sand flats occur below low tide level. These are generally covered with large ripples to small sand dunes. Between the low tide and high tide marks, ripples are abundant on a mixed sand and mud flat. A mud flat, backed by a tidal marsh, forms above the high tide mark. Landward of the low tide level, tidal creeks cut through the deposits as well.

Marine Environments

Sediments may accumulate and be preserved virtually anywhere in the oceans. Consequently, marine sedimentary environments are numerous and widespread. Water depth also plays a major role in shaping these environments. For simplicity, marine environments can be divided into two broad groups: shelf and deep oceanic. Shelf environments range in depth from low tide level to depths of 425 ft (130 m), typical for the outer edge of the continental shelf.

Continental Shelf Environments

The average continental shelf is about 45 mi (75 km) wide. Shelf sediments generally decrease in grain size with increasing distance from shore. This occurs for two reasons:

1. Greater distance from sediment sources and
2. Decreasing sediment movement (transport) with increasing water depth.

Shelf sediments vary significantly with latitude. At high latitudes, glacial ice flowing into coastal water generates icebergs, which transport large sediment loads of various sizes out onto the shelf. As icebergs melt, they drop their load. These glaciomarine sediments are generally less sorted and coarser grained than lower latitude deposits. In fact, boulders known as dropstones occur on the sea floor in deep water, hundreds of miles from shore.

Rivers deliver most of the sediments to mid-latitude shelves. Therefore, grain size routinely decreases with distance from shore; sediment sorting also tends to be good. Shallow water, nearshore sediments form thick sand blankets with abundant ripple marks. As depth increases and water movement decreases, average grain size decreases, and sand, silt, and clay occur interbedded. In water depths greater than 150 to 200 ft (45-60 m), even storm waves do not stir the bottom; consequently, silts and clays predominate. Scattered sand deposits are also located on outer shelf margins. During periods of lower sea level, rivers flowing across what is now the inner shelf deposited these so-called relict sediments.

At low latitudes, bottom-dwelling plants and animals secrete large volumes of calcium carbonate, producing thick blankets of carbonate sediment. Perhaps the best known carbonate environment is the coral reef. Corals produce a rigid framework of carbonate rock (limestone), which is also a major source of sediment of various grain sizes. Where stream input is great, terrigenous sediments discourage habitation by carbonate-producing organisms and dilute any carbonate sediment that is produced.

Deep Oceanic Environments

Seaward of the continental shelves, continental slopes incline more steeply, so relict and modern sediments form deposits called deep-sea fans. These are similar to alluvial fans, but generally consist of sand-to clay-sized particles with little or no gravel. Deep-sea fans form the continental rise, a continuous apron of sediment at the base of the continental slope.

Even farther from land, the monotonous abyssal plains begin. Here mostly clay-sized sediment forms sheets up to 0.6 mi (1 km) thick. These deposits, composed of sediments that settle through the water column from shallow depths, thin to a feather edge at the oceanic ridges where new sea floor forms. Abyssal sediments are generally a mixture of three grain types: carbonate muds and siliceous muds of biogenic (organic) origin, and red clays of terrigenous origin. Carbonate-rich muds generally accumulate in water depths of less than 2 to 2.5 mi (3-4 km); at deeper depths, colder water and higher pressures combine to dissolve the carbonate. Siliceous muds occur where abundant nutrients in surface waters support high rates of biogenic silica (SiO_2) production. Red clays, transported from the land by winds and stream flow, predominate where quantities of carbonate and siliceous muds are insufficient to dilute these finegrained terrigenous deposits.

Types of Sedimentary Rocks

Breccia

Breccia is a sedimentary rock made up of angular particles over two millimeters in diameter (clasts) with the spaces between the particles filled with smaller particles and mineral cement (matrix). The word "breccia" has an Italian origin, meaning "stone made of cemented gravel." The rock occurs worldwide and has been found on the Moon and Mars.

Formation of Breccia

Like other clastic sedimentary rocks, breccia forms when other rock is subjected to weathering. The clasts are angular and irregular, indicating the particles forming the rock didn't travel far from their source. Other material fills in the spaces between the clasts, binding them into rock. One way to categorize breccia is by its method of formation. For example:

- Some breccia forms as material accumulates at the base of a steep slope or cliff.
- Cataclastic breccia is formed when fragments fall from a fault.
- Volcanic breccia, pyroclastic, or igneous breccia is formed from compaction of lava chunks with ash.
- Collapse breccia is sedimentary breccia formed from the collapse of a cavern.
- Impact breccia is formed from a meteor impact breaking rock at the impact site.
- Hydrothermal breccia is formed when fluid fractures a rock.

The spaces between the clasts fill with silt (iron oxide), carbonate (e.g., calcite), or silica, eventually acting as a cement that binds the particles.

Sometimes the deposition of clast and matrix material occurs about the same time. Another class of breccia consists of rock in which the clasts and matrix are unrelated. For example, the collapse of a limestone cavern would produce both clasts and matrix material at once, while a mudslide over a fault would coat old clastic material with young matrix.

Another way to classify breccia is by the distribution of clasts and matrix. In matrix supported breccia, clasts don't touch each other and matrix completely surrounds them. In clast supported breccia, matrix fills the void between touching (or nearly continuous) clasts.

Breccia Composition

This is an example of matrix supported breccia.

Breccia usually refers to rock of sedimentary origin, although it may also form from igneous or metamorphic rocks. A mixture of different rocks and minerals may combine.

Thus, breccia composition and properties are highly variable. Usually, clasts consist of a hard, durable rock that can survive some degree of weathering. Sometimes breccia is named to reference its composition. For example, there is sandstone breccia, basalt breccia, and chert breccia. Monomict breccia is breccia containing clasts of a single rock type. Polymict breccia or petromict breccia is breccia containing clasts of different rocks.

Properties

Historically, breccia has been used as a building material.

The identifying feature of breccia is that it consists of visible angular clasts cemented together with another mineral. The clasts should be easily visible to the naked eye. Otherwise, the properties of the rock are highly variable. It can occur in any color, and may be either hard or soft. The rock may be rough to the touch because of the angular clasts. Whether it polishes to a smooth surface depends on the similarity of clast and matrix composition.

Breccia uses

Breccia may be polished to make gems and decorative items.

Because of its variable composition, breccia has an interesting appearance. The rock is mainly used to make sculptures, gems, and architectural elements. The Minoan palace of Knossos on Crete, constructed around 1800 BC, includes columns made of breccia. The ancient Egyptians used breccia to make statues. The Romans regarded breccia as a precious stone and used it to construct public buildings, columns, and walls. The Pantheon in Rome features columns made of pavonazzetto, a type of breccia with a pattern resembling peacock feathers. In modern culture, breccia is used for decorative elements, jewelry, and sometimes as a fill material for roads.

Breccia versus Conglomerate

The clasts in conglomerate rock are rounded. In breccia, they are irregularly-shaped.

Breccia and congomerate are similar to each other. Both are clastic sedimentary rocks containing clasts larger than two millimeters in diameter. The difference is that the clasts in breccia are angular, while those in conglomerate are rounded. This indicates the clasts in conglomerate traveled a greater distance from their source or experienced more weathering before becoming embedded in matrix than the clasts in breccia.

Conglomerate

Conglomerate is a clastic sedimentary rock that is formed by the accumulation of rock fragments greater than 2 millimetre in size. The rock fragments that forms the conglomerate are round in shape. The pore spaces in between the clast of a rock are filled by the finer particles usually silt and clay or any other cementing material that binds the rock fragments together.

Conglomerate can be formed of any rock material that is transported through time and space from the origin area to accumulation area. The rounded rock fragments is the proof of transportation of the rock fragments in the conglomerate. The rock fragments of the conglomerate can be derived from igneous, metamorphic or sedimentary rocks. The chemical binding material can be sand, silt or chemical cement like calcite.

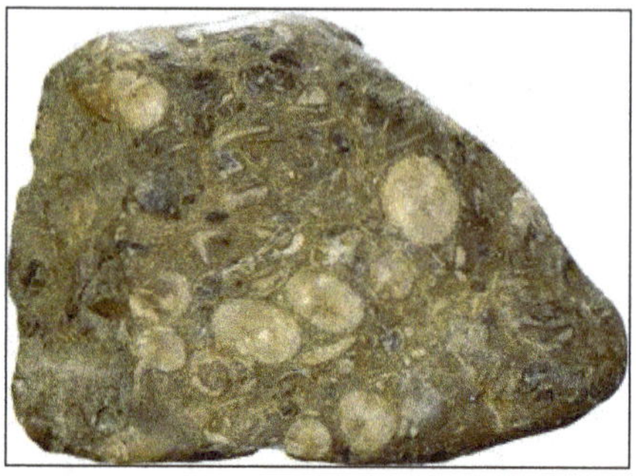

Composition of Conglomerate

Conglomerate can have a variety of compositions. As a clastic sedimentary rock, it can contain clasts of any rock material or weathering product that is washed downstream or down current. The rounded clasts of conglomerate can be mineral particles such as quartz, or they can be sedimentary, metamorphic, or igneous rock fragments. The matrix that binds the large clasts together can be a mixture of sand, mud, and chemical cement.

Generation of Conglomerate

Conglomerate can be formed at an areas where strong water current exist like mountain down slope where water has enough current flow that it can carry the rock fragments above 2 millimetre. It can also be formed at beaches where water current is strong and rock fragments are available to be accumulated for forming conglomerate. Conglomerate is formed when large clast pebble or cobble size fragments transported and deposited than the finer grained fills the spaces in between the clast.

Classification of Conglomerate

Conglomerates may be named and classified by the:

- Amount and type of matrix present.
- Composition of gravel-size clasts they contain.
- Size range of gravel-size clasts present.

A sedimentary rock composed largely of gravel is first named according to the roundness of the gravel. If the gravel clasts that comprise it is largely well-rounded to subrounded, it is a conglomerate. If the gravel clasts that comprise it are largely angular, it is a breccia. Such breccias can be called sedimentary breccias to differentiate them from other types of breccia, e.g. volcanic and fault breccias. Sedimentary rocks that contain a mixture of rounded and angular gravel clasts are sometimes called breccio-conglomerate.

Texture

Conglomerates are rarely composed entirely of gravel-size clasts. Typically, the space between the gravel-size clasts is filled by a mixture composed of varying amounts of silt, sand, and clay, known as matrix. If the individual gravel clasts in a conglomerate are separated from each other by an abundance of matrix such that they are not in contact with each other and float within the matrix, it is called a paraconglomerate. Paraconglomerates are also often unstratified and can contain more matrix than gravel clasts. If the gravel clasts of a conglomerate are in contact with each other, it is called a orthoconglomerate. Unlike paraconglomerates, orthoconglomerates are typically cross-bedded and often well-cemented and lithified by either calcite, hematite, quartz, or clay.

The differences between paraconglomerates and orthoconglomerates reflect differences in how they are deposited. Paraconglomerates are commonly either glacial tills or debris flow deposits. Orthoconglomerates are tyipically associated with aqueous currents of some sort.

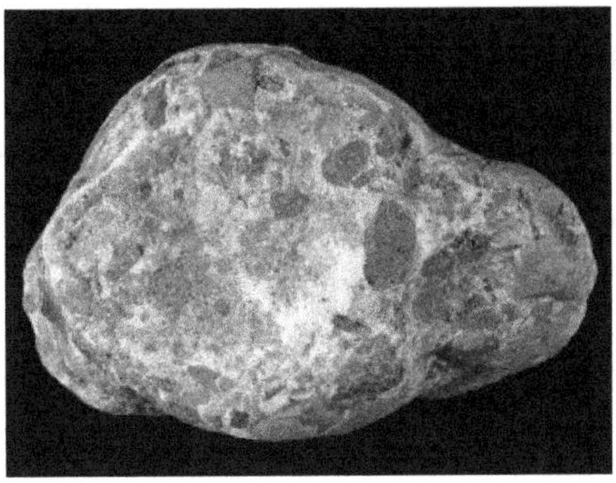

Clast Composition

Conglomerates are also classified according to the composition of their clasts. A conglomerate or any clastic sedimentary rock that consists of a single rock or mineral is known as either a monomict, monomictic, oligomict, or oligomictic conglomerate. If the conglomerate consists of two or more different types of rocks, minerals, or combination of both, it is known as either a polymict or polymictic conglomerate. If a polymictic conglomerate contains an assortment of the clasts of metastable and unstable rocks and minerals, it called either a petromict or petromictic conglomerate.

In addition, conglomerates are classified by source as indicated by the lithology of the gravel-size clasts If these clasts consist of rocks and minerals that are significantly different in lithology from the enclosing matrix and, thus, older and derived from outside the basin of deposition, the conglomerate is known as an extraformational conglomerate. If these clasts consist of rocks and minerals that are identical to or consistent with the lithology of the enclosing matrix and, thus, penecontemporaneous and derived from within the basin of deposition, the conglomerate is known as an intra-formational conglomerate.

Two recognized types of type of intra-formational conglomerates are shale-pebble and flat-pebble conglomerates. A shale-pebble conglomerate is a conglomerate that is composed largely of clasts of rounded mud chips and pebbles held together by clay minerals and created by erosion within environments such as within a river channel or along a lake margin. Flat-pebble conglomerates (edgewise conglomerates) are conglomerates that consist of relatively flat clasts of lime mud created by either storms or tsunami eroding a shallow sea bottom or tidal currents eroding tidal flats along a shoreline.

Clast Size

Finally, conglomerates are often differentiated and named according to the dominant clast size comprising them. In this classification, a conglomerate composed largely of granule-size clasts would be called a granule conglomerate; a conglomerate composed largely of pebble-size clasts would be called a pebble conglomerate; and a conglomerate composed largely of cobble-size clasts would be called a cobble conglomerate.

Conglomerate Forming Sedimentary Environments

Conglomerates are deposited in a variety of sedimentary environments.

1. Deepwater marine: In turbidites, the basal part of a bed is typically coarse-grained and sometimes conglomeratic. In this setting, conglomerates are normally very well sorted, well-rounded and often with a strong A-axis type imbrication of the clasts.
2. Shallow marine: Conglomerates are normally present at the base of sequences laid down during marine transgressions above an unconformity, and are known as basal conglomerates. They represent the position of the shoreline at a particular time and are diachronous.
3. Fluvial: Conglomerates deposited in fluvial environments are typically well rounded and well sorted. Clasts of this size are carried as bedload and only at times of high flow-rate. The maximum clast size decreases as the clasts are transported further due to attrition, so conglomerates are more characteristic of immature river systems. In the sediments deposited by mature rivers, conglomerates are generally confined to the basal part of a channel fill where they are

known as pebble lags. Conglomerates deposited in a fluvial environment often have an AB-plane type imbrication.

4. Alluvial: Alluvial deposits form in areas of high relief and are typically coarse-grained. At mountain fronts individual alluvial fans merge to form braidplains and these two environments are associated with the thickest deposits of conglomerates. The bulk of conglomerates deposited in this setting are clast-supported with a strong AB-plane imbrication. Matrix-supported conglomerates, as a result of debris-flow deposition, are quite commonly associated with many alluvial fans. When such conglomerates accumulate within an alluvial fan, in rapidly eroding (e.g., desert) environments, the resulting rock unit is often called a fanglomerate.
5. Glacial: Glaciers carry a lot of coarse-grained material and many glacial deposits are conglomeratic. Tillites, the sediments deposited directly by a glacier, are typically poorly sorted, matrix-supported conglomerates. The matrix is generally fine-grained, consisting of finely milled rock fragments. Waterlaid deposits associated with glaciers are often conglomeratic, forming structures such as eskers.

Examples of Conglomerate

An example of conglomerate can be seen at Montserrat, near Barcelona. Here, erosion has created vertical channels that give the characteristic jagged shapes the mountain is named for (Montserrat literally means "jagged mountain"). The rock is strong enough to use as a building material, as in the Santa Maria de Montserrat Abbey.

Another example, the Crestone Conglomerate, occurs in and near the town of Crestone, at the foot of the Sangre de Cristo Range in Colorado's San Luis Valley. The Crestone Conglomerate consists of poorly sorted fanglomerates that accumulated in prehistoric alluvial fans and related fluvial systems. Some of these rocks have hues of red and green.

It shows an outcrop of conglomerate and some pebble-size weathering debris. The round pebbles are too large to have been moved and shaped by wind, thus they had to have been transported a significant distance by water.

Conglomerate cliffs are found on the east coast of Scotland from Arbroath northwards along the coastlines of the former counties of Angus and Kincardineshire. Dunottar Castle sits on a rugged promontory of conglomerate jutting into the North Sea just south of the town of Stonehaven.

Conglomerate may also be seen in the domed hills of Kata Tjuta, in Australia's Northern Territory.

In the nineteenth century a thick layer of Pottsville conglomerate was recognised to underlie anthracite coal measures in Pennsylvania.

Uses of Conglomerate

Conglomerate has very few uses because of it not clean breakage and fine particles are unreliable. It can only be used as a crush where low performance material is wanted. Conglomerate has very few commercial uses. Its inability to break cleanly makes it a poor candidate for dimension stone, and its variable composition makes it a rock of unreliable physical strength and durability. Conglomerate can be crushed to make a fine aggregate that can be used where a low-performance material is suitable. Many conglomerates are colorful and attractive rocks, but they are only rarely used as an ornamental stone for interior use.

Analysis of conglomerate can sometimes be used as a prospecting tool. For example, most diamond deposits are hosted in kimberlite. If a conglomerate contains clasts of kimberlite, then the source of that kimberlite must be somewhere upstream.

Siltstone

Siltstone is a sedimentary rock composed mainly of silt-sized particles. It forms where water, wind, or ice deposit silt, and the silt is then compacted and cemented into a rock.

Siltstone Colors: Siltstone occurs in a wide variety of colors. It is usually gray, brown, or reddish brown. It can also be white, yellow, green, red, purple, orange, black, and other colors. The colors are a response to the composition of the grains, the composition of the cement, or stains from subsurface waters. Specimens in the photo are about two inches across.

Silt accumulates in sedimentary basins throughout the world. It represents a level of current, wave, or wind energy between where sand and mud accumulate. These include fluvial, aeolian, tidal, coastal, lacustrine, deltaic, glacial, paludal, and shelf environments. Sedimentary structures such as layering, cross-bedding, ripple marks, erosional contacts, and fossils provide evidence of these environments.

Siltstone is much less common than sandstone and shale. The rock units are usually thinner and less extensive. Only rarely is one notable enough to merit a stratigraphic name.

Silt

The word "silt" does not refer to a specific substance. Instead, it is a word used for loose granular particles in a specific size range.

Silt-sized particles range between 0.00015 and 0.0025 inches in diameter, or between 0.0039 and 0.063 millimeters in diameter. They are intermediate in size between coarse clay on the small side and fine sand on the large side.

Grains of coarse silt are large enough that most people can see them without magnification on a background of contrasting color. Most people are not able to sense them if they roll a few grains of silt between their thumb and index finger. Most people are able to detect a few grains of silt by biting them gently between their front teeth. (This test is not recommended, but some experienced geologists and soil scientists use it for quick field identification of silt in sediment and soil.)

Silt does not have a definite composition. It is usually a mixture of clay minerals, micas, feldspars, and quartz. The small-size fraction of silt is mostly clay. The coarse-size fraction is mostly grains of feldspar and quartz.

Siltstone Outcrop: An exposure of the Holtzclaw siltstone near Louisville, Kentucky. It shows the thinly bedded and differentially weathered character of the rock unit. Siltstones are rarely of sufficient thickness or lateral persistence to merit a stratigraphic name.

Color of Siltstone

Siltstone occurs in a wide range of colors. It is usually gray, brown, or reddish brown. White, yellow, green, red, purple, orange, black, and other colors occur. The color is caused by the composition of the grains, the composition of the cement that binds them together, and stains produced by contact with subsurface waters.

Field Identification

Siltstone can be difficult to identify in the field without close examination. Weathered surfaces often appear to show sedimentary structures where none are present. Different layers weather at different rates. Siltstone is often interbedded with other lithologies.

Identification requires breaking off a small piece and observing the grain size. Scraping the surface with a nail or knife blade will dislodge tiny silt grains instead of dislodging sand grains or producing a white effervescent powder.

Siltstone uses and Economics

Siltstone has very few uses. It is rarely the target of mining for use as a construction material or manufacturing feedstock. The intergranular pore spaces in siltstone are too small for it to serve as a good aquifer. It is rarely porous enough or extensive enough to serve as an oil or gas reservoir. Its main use is as a low-quality fill when better materials are not locally available.

Shale

Shale is a fine-grained sedimentary rock that forms from the compaction of silt and clay-size mineral particles that we commonly call “mud.” This composition places shale in a category of sedimentary rocks known as “mudstones.” Shale is distinguished from other mudstones because it is fissile and laminated. “Laminated” means that the rock is made up of many thin layers. “Fissile” means that the rock readily splits into thin pieces along the laminations.

Shale: Shale breaks into thin pieces with sharp edges. It occurs in a wide range of colors that include red, brown, green, gray, and black. It is the most common sedimentary rock and is found in sedimentary basins worldwide.

Uses of Shale

Some shales have special properties that make them important resources. Black shales contain organic material that sometimes breaks down to form natural gas or oil. Other shales can be crushed and mixed with water to produce clays that can be made into a variety of useful objects.

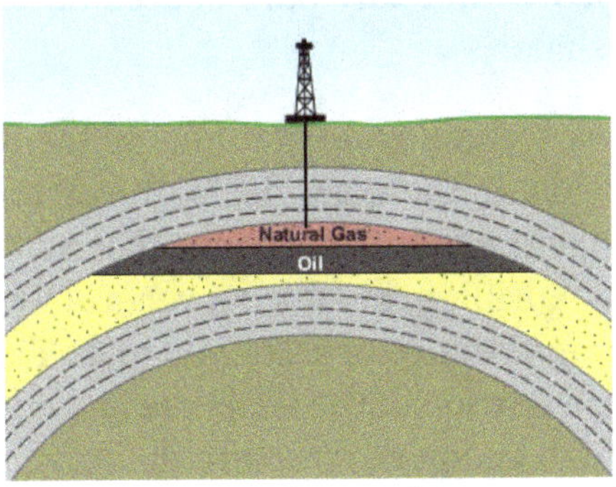

Conventional Oil and Natural Gas Reservoir.

In figure, illustrated an "anticlinal trap" that contains oil and natural gas. The gray rock units are impermeable shale. Oil and natural gas forms within these shale units and then migrates upwards. Some of the oil and gas becomes trapped in the yellow sandstone to form an oil and gas reservoir. This is a "conventional" reservoir - meaning that the oil and gas can flow through the pore space of the sandstone and be produced from the well.

Conventional Oil and Natural Gas

Black organic shales are the source rock for many of the world's most important oil and natural gas deposits. These shales obtain their black color from tiny particles of organic matter that were deposited with the mud from which the shale formed. As the mud was buried and warmed within the earth, some of the organic material was transformed into oil and natural gas.

The oil and natural gas migrated out of the shale and upwards through the sediment mass because of their low density. The oil and gas were often trapped within the pore spaces of an overlying rock unit such as sandstone. These types of oil and gas deposits are known as "conventional reservoirs" because the fluids can easily flow through the pores of the rock and into the extraction well.

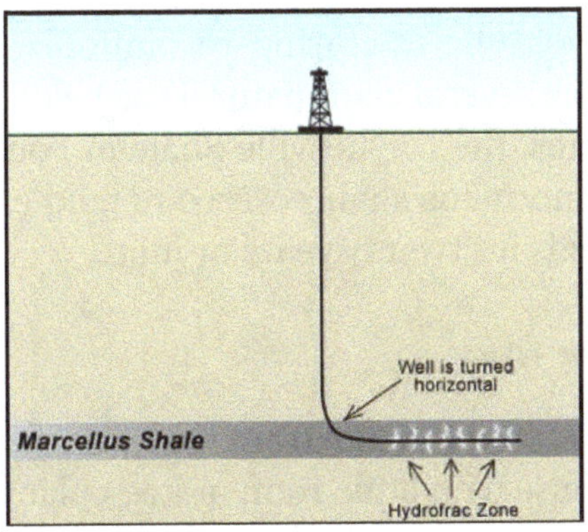

Unconventional Oil and Gas Reservoir.

Although drilling can extract large amounts of oil and natural gas from the reservoir rock, much of it remains trapped within the shale. This oil and gas is very difficult to remove because it is trapped within tiny pore spaces or adsorbed onto clay mineral particles that make up the shale.

In figure, illustrates the new technologies that enable the development of unconventional oil and natural gas fields. In these gas fields, the oil and gas are held in shales or another rock unit that is impermeable. To produce that oil or gas, special technologies are needed. One is horizontal drilling, in which a vertical well is deviated to horizontal so that it will penetrate a long distance of reservoir rock. The second is hydraulic fracturing. With this technique, a portion of the well is sealed off and water is pumped in to produce a pressure that is high enough to fracture the surrounding rock. The result is a highly fractured reservoir penetrated by a long length of well bore.

Unconventional Oil and Natural Gas

In the late 1990s, natural gas drilling companies developed new methods for liberating oil and natural gas that is trapped within the tiny pore spaces of shale. This discovery was significant because it unlocked some of the largest natural gas deposits in the world.

The Barnett Shale of Texas was the first major natural gas field developed in a shale reservoir rock. Producing gas from the Barnett Shale was a challenge. The pore spaces in shale are so tiny that the gas has difficulty moving through the shale and into the well. Drillers discovered that they could increase the permeability of the shale by pumping water down the well under pressure that was high enough to fracture the shale. These fractures liberated some of the gas from the pore spaces and allowed that gas to flow to the well. This technique is known as “hydraulic fracturing” or “hydrofracing.”

Drillers also learned how to drill down to the level of the shale and turn the well 90 degrees to drill horizontally through the shale rock unit. This produced a well with a very long “pay zone” through the reservoir rock. This method is known as “horizontal drilling.”

Horizontal drilling and hydraulic fracturing revolutionized drilling technology and paved the way for developing several giant natural gas fields. These include the Marcellus Shale in the Appalachians, the Haynesville Shale in Louisiana and the Fayetteville Shale in Arkansas. These enormous shale reservoirs hold enough natural gas to serve all of the United States’ needs for twenty years or more.

Shale used to Produce Clay

Everyone has contact with products made from shale. If you live in a brick house, drive on a brick road, live in a house with a tile roof, or keep plants in “terra cotta” pots, you have daily contact with items that were probably made from shale.

Many years ago these same items were made from natural clay. However, heavy use depleted most of the small clay deposits. Needing a new source of raw materials, manufacturers soon discovered that mixing finely ground shale with water would produce clay that often had similar or superior properties. Today, most items that were once produced from natural clay have been replaced by almost identical items made from clay manufactured by mixing finely ground shale with water.

Shale used to Produce Cement

Cement is another common material that is often made with shale. To make cement, crushed limestone and shale are heated to a temperature that is high enough to evaporate off all water and break down the limestone into calcium oxide and carbon dioxide. The carbon dioxide is lost as an emission, but the calcium oxide combined with the heated shale makes a powder that will harden if mixed with water and allowed to dry. Cement is used to make concrete and many other products for the construction industry.

Shale in brick and tile: Shale is used as a raw material for making many types of brick, tile, pipe, pottery, and other manufactured products. Brick and tile are some of the most extensively used and highly desired materials for building homes, walls, streets, and commercial structures.

Oil Shale

Oil shale: A rock that contains a significant amount of organic material in the form of solid kerogen. Up to 1/3 of the rock can be solid organic material. This specimen is approximately four inches (ten centimeters) across.

Oil shale is a rock that contains significant amounts of organic material in the form of kerogen. Up to 1/3 of the rock can be solid kerogen. Liquid and gaseous hydrocarbons can be extracted from oil shale, but the rock must be heated and treated with solvents. This is usually much less efficient than drilling rocks that will yield oil or gas directly into a well. Extracting the hydrocarbons from oil shale produces emissions and waste products that cause significant environmental concerns. This is one reason why the world's extensive oil shale deposits have not been aggressively utilized.

Oil shale usually meets the definition of "shale" in that it is "a laminated rock consisting of at least 67% clay minerals." However, it sometimes contains enough organic material and carbonate minerals that clay minerals account for less than 67% of the rock.

Composition of Shale

Shale is a rock composed mainly of clay-size mineral grains. These tiny grains are usually clay minerals such as illite, kaolinite, and smectite. Shale usually contains other clay-size mineral particles such asquartz, chert, and feldspar. Other constituents might include organic particles, carbonate minerals, iron oxide minerals, sulfide minerals, and heavy mineral grains. These "other constituents" in the rock are often determined by the shale's environment of deposition, and they often determine the color of the rock.

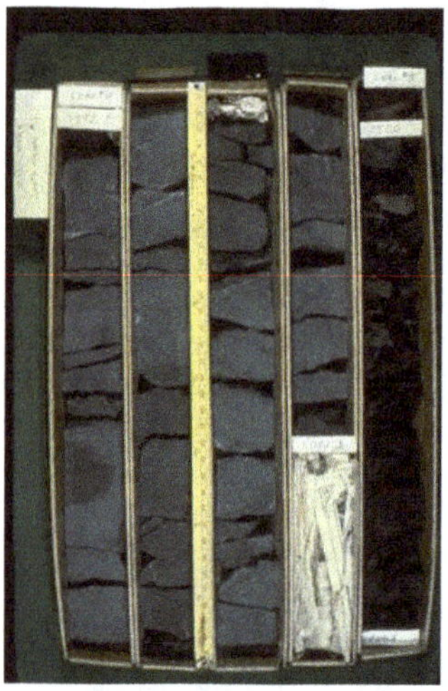

Shale core samples: When shale is drilled for oil, natural gas, or mineral resource evaluation, a core is often recovered from the well.

Colors of Shale

Like most rocks, the color of shale is often determined by the presence of specific materials in minor amounts. Just a few percent of organic materials or iron can significantly alter the color of a rock.

Black and Gray Shale

A black color in sedimentary rocks almost always indicates the presence of organic materials. Just one or two percent organic materials can impart a dark gray or black color to the rock. In addition, this black color almost always implies that the shale formed from sediment deposited in an oxygen-deficient environment. Any oxygen that entered the environment quickly reacted with the decaying organic debris. If a large amount of oxygen was present, the organic debris would all have decayed. An oxygen-poor environment also provides the proper conditions for the formation of sulfide minerals such as pyrite, another important mineral found in most black shales.

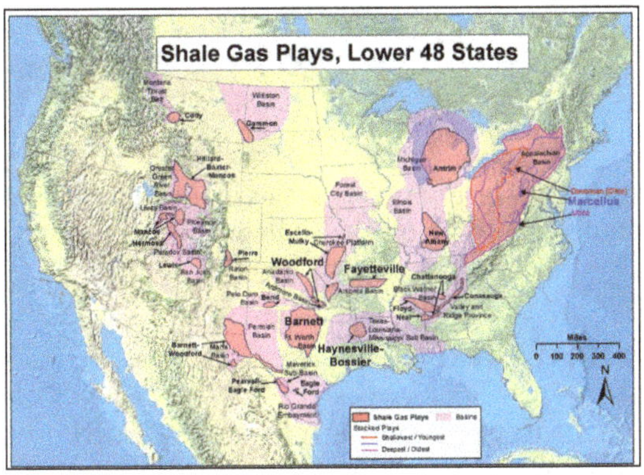

Shale gas plays: Since the late 1990s, dozens of previously unproductive black organic shales have been successfully developed into valuable gas fields.

The presence of organic debris in black shales makes them the candidates for oil and gas generation. If the organic material is preserved and properly heated after burial, oil and natural gas might be produced. The Barnett Shale, Marcellus Shale, Haynesville Shale, Fayetteville Shale, and other gas-producing rocks are all dark gray or black shales that yield natural gas. The Bakken Shale of North Dakota and the Eagle Ford Shale of Texas are examples of shales that yield oil.

Black shale: Organic-rich black shale. Natural gas and oil are sometimes trapped in the tiny pore spaces of this type of shale.

Gray shales sometimes contain a small amount of organic matter. However, gray shales can also be rocks that contain calcareous materials or simply clay minerals that result in a gray color.

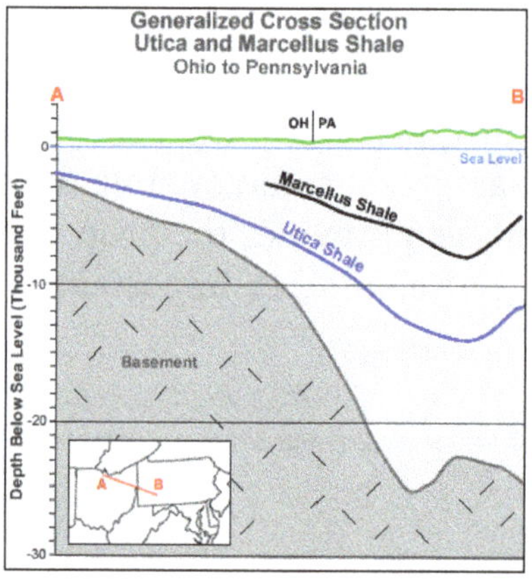

Utica and Marcellus Shale: Two black organic shales in the Appalachian Basin are thought to contain enough natural gas to supply the United States for several years. These are the Marcellus Shale and Utica Shale.

Red, Brown and Yellow Shale

Shales that are deposited in oxygen-rich environments often contain tiny particles of iron oxide or iron hydroxide minerals such as hematite, goethite, or limonite. Just a few percent of these minerals distributed through the rock can produce the red, brown, or yellow colors exhibited by many types of shale. The presence of hematite can produce red shale. The presence of limonite or goethite can produce yellow or brown shale.

Green Shale

Green shales are occasionally found. This should not be surprising because some of the clay minerals and micas that make up much of the volume of these rocks are typically a greenish color.

Hydraulic Properties of Shale

Hydraulic properties are characteristics of a rock such as permeability and porosity that reflect its ability to hold and transmit fluids such as water, oil, or natural gas.

Natural gas shale well: In less than ten years, shale has skyrocketed to prominence in the energy sector. New drilling and well development methods such as hydraulic fracturing and horizontal drilling can tap the oil and natural gas trapped within the tight matrix of organic shales.

Shale has a very small particle size, so the interstitial spaces are very small. In fact they are so small that oil, natural gas, and water have difficulty moving through the rock. Shale can therefore serve as a cap rock for oil and natural gas traps, and it also is an aquiclude that blocks or limits the flow of groundwater.

Although the interstitial spaces in a shale are very small, they can take up a significant volume of the rock. This allows the shale to hold significant amounts of water, gas, or oil but not be able to effectively transmit them because of the low permeability. The oil and gas industry overcomes these limitations of shale by using horizontal drilling and hydraulic fracturing to create artificial porosity and permeability within the rock.

Some of the clay minerals that occur in shale have the ability to absorb or adsorb large amounts of water, natural gas, ions, or other substances. This property of shale can enable it to selectively and tenaciously hold or freely release fluids or ions.

Engineering Properties of Shale Soils

Shales and the soils derived from them are some of the most troublesome materials to build upon. They are subject to changes in volume and competence that generally make them unreliable construction substrates.

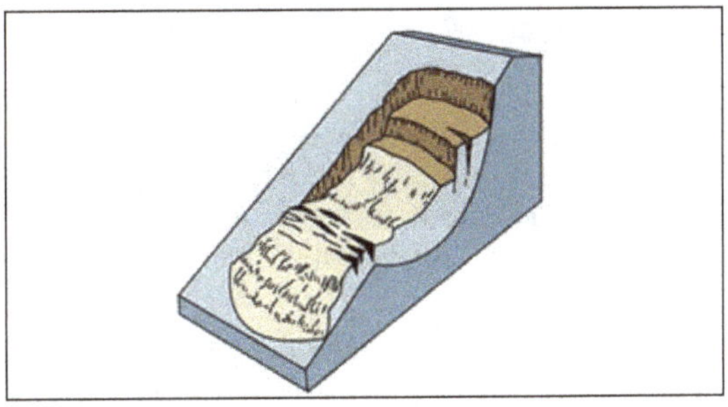

Landslide: Shale is a landslide-prone rock.

Expansive Soils

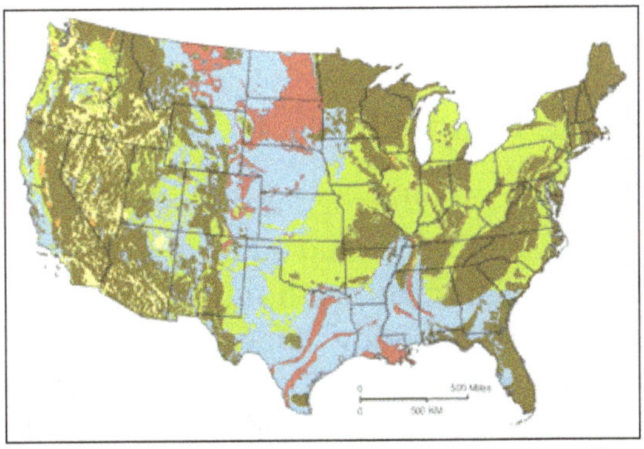

Expansive soils map: The United States Geological Survey has prepared a generalized expansive soils mapfor the lower 48 states.

The clay minerals in some shale-derived soils have the ability to absorb and release large amounts of water. This change in moisture content is usually accompanied by a change in volume which can be as much as several percent. These materials are called "expansive soils." When these soils become wet they swell, and when they dry out they shrink. Buildings, roads, utility lines, or other structures placed upon or within these materials can be weakened or damaged by the forces and motion of volume change. Expansive soils are one of the most common causes of foundation damage to buildings in the United States.

Slope Stability

Shale is the rock most often associated with landslides. Weathering transforms the shale into a clay-rich soil which normally has a very low shear strength - especially when wet. When these low-strength materials are wet and on a steep hillside, they can slowly or rapidly move down slope. Overloading or excavation by humans will often trigger failure.

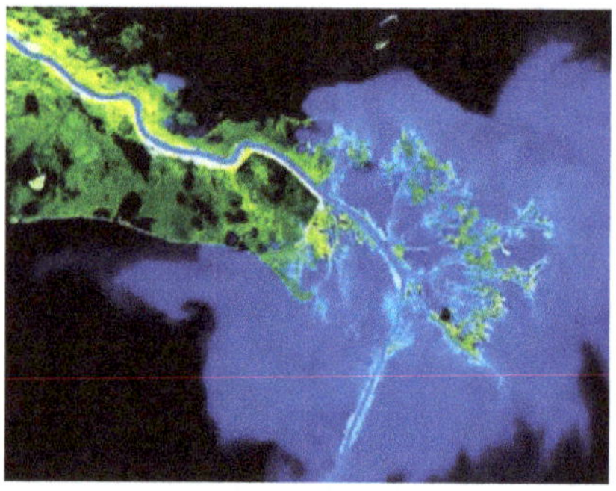

Shale delta.

In figure, a delta is a sediment deposit that forms when a stream enters a standing body of water. The water velocity of the stream suddenly decreases and the sediments being carried settle to the bottom. Deltas are where the largest volume of Earth's mud is deposited. The image above is a satellite view of the Mississippi delta, showing its distributary channels and inter-distributary deposits. The bright blue water surrounding the delta is laden with sediment.

Environments of Shale Deposition

An accumulation of mud begins with the chemical weathering of rocks. This weathering breaks the rocks down into clay minerals and other small particles which often become part of the local soil. A rainstorm might wash tiny particles of soil from the land and into streams, giving the streams a "muddy" appearance. When the stream slows down or enters a standing body of water such as a lake, swamp, or ocean, the mud particles settle to the bottom. If undisturbed and buried, this accumulation of mud might be transformed into a sedimentary rock known as "mudstone." This is how most shales are formed.

Shale on Mars: Shale is also a very common rock on Mars. It shows thinly bedded fissile shales outcropping in the Gale Crater. Curiosity drilled holes into the rocks of Gale Crater and identified clay minerals in the cuttings.

The shale-forming process is not confined to Earth. The Mars rovers have found lots of outcrops on Mars with sedimentary rock units that look just like the shales found on Earth.

Sandstone

Sandstone is a sedimentary rock composed mostly of quartz sand, but it can also contain significant amounts of feldspar, and sometimes silt and clay. Sandstone that contains more than 90% quartz is called quartzose sandstone. When the sandstone contains more than 25% feldspar, it is called arkose or arkosic sandstone. When there is a significant amount of clay or silt, geologists refer to the rock as argillaceous sandstone. The color of sandstone varies, depending on its composition. Argillaceous sandstones are often gray to blue. Because it is composed of light colored minerals, sandstone is typically light tan in color. Other elements, however, create colors in sandstone. The most common sandstones have various shades of red, caused by iron oxide (rust). In some instances, there is a purple hue caused by manganese.

Rock

Sandstone is a sedimentary rock composed mostly of quartz sand, but it can also contain significant amounts of feldspar, and sometimes silt and clay. Sandstone that contains more than 90% quartz is called quartzose sandstone. When the sandstone contains more than 25% feldspar, it is called arkose or arkosic sandstone. When there is a significant amount of clay or silt, geologists refer to the rock as argillaceous sandstone. The color of sandstone varies, depending on its composition. Argillaceous sandstones are often gray to blue. Because it is composed of light colored minerals, sandstone is typically light tan in color. Other elements, however, create colors in sandstone. The most common sandstones have various shades of red, caused by iron oxide (rust). In some instances, there is a purple hue caused by manganese.

Relation to Mining

Loosely consolidated material can be mined using a high-pressure hydraulic monitor. Sand washed from the working face is collected in a sump, from which it is usually pumped to a dewatering/surge pile, where the material is allowed to free drain. Once drained, the sand reports to a plant.

Well-consolidated deposits can be mined using conventional drill-and-blast methods. Once blasted, the sandstone is "mucked" using a front-end loader, placed into haul trucks, and transported to a primary crusher. Following crushing, the material may report to a secondary or tertiary crushing plant. The sized material is then transported to a processing plant, usually by a conveyance system.

Uses

Sandstone has two major applications, as crushed stone and as dimension stone.

Dimension stone is any rock material that is cut into specific sizes, typically as blocks and slabs. Crushed stone is used in the construction of roadways and road structures such as bridges, and in buildings, both commercial and residential.

Iron Ore

Iron Ore: A specimen of oolitic hematite iron ore. The specimen shown is about two inches (five centimeters) across.

Earth's most important iron ore deposits are found in sedimentary rocks. They formed from chemical reactions that combined iron and oxygen in marine and fresh waters. The two most important minerals in these deposits are iron oxides: hematite (Fe_2O_3) and magnetite (Fe_3O_4). These iron ores have been mined to produce almost every iron and steel object that we use today - from paper clips to automobiles to the steel beams in skyscrapers.

Formation of Iron Ore

Nearly all of Earth's major iron ore deposits are in rocks that formed over 1.8 billion years ago. At that time Earth's oceans contained abundant dissolved iron and almost no dissolved oxygen. The iron ore deposits began forming when the first organisms capable of photosynthesis began releasing oxygen into the waters. This oxygen immediately combined with the abundant dissolved iron to produce hematite or magnetite. These minerals deposited on the sea floor in great abundance, forming what are now known as the "banded iron formations." The rocks are "banded" because the iron minerals deposited in alternating bands with silica and sometimes shale. The banding might have resulted from seasonal changes in organism activity.

Steel Mill: Most iron ore is used to make steel. Here a steel slab is being cut to length in a steel mill.

Uses of Iron Ore

Banded Iron Formation: Close-up view of a banded iron formation. In this specimen bands of hematite (silver) alternate with bands of jasper (red). This photo spans an area of rock about one foot wide.

The primary use of iron ore is in the production of iron. Most of the iron produced is then used to make steel. Steel is used to make automobiles, locomotives, ships, beams used in buildings, furniture, paper clips, tools, reinforcing rods for concrete, bicycles, and thousands of other items. It is the most-used metal by both tonnage and purpose.

Chert

Chert is a fine-grained, silica-rich, microcrystalline, cryptocrystalline or microfibrous sedimentary rock that may contain small fossils. It varies greatly in color (from white to black), but most often manifests as gray, brown, grayish brown and light green to rusty red; its color is an expression of trace elements present in the rock, and both red and green are most often related to traces of iron (in its oxidized and reduced forms, respectively).

Chert.

Chert occurs as oval to irregular nodules in greensand, limestone, chalk, and dolostone formations as a replacement mineral, where it is formed as a result of some type of diagenesis. Where it occurs in chalk, it is usually called flint. It also occurs in thin beds, when it is a primary deposit (such as with many jaspers and radiolarites). Thick beds of chert occur in deep geosynclinal deposits. These thickly bedded cherts include the novaculite of the Ouachita Mountains of Arkansas, Oklahoma, and similar occurrences in Texas in the United States. The banded iron formations of Precambrian age are composed of alternating layers of chert and iron oxides.

Concerning the Terms “Chert,” “Chalcedony” and “Flint”

There is much confusion concerning the exact meanings and differences among the terms “chert,” “chalcedony” and “flint” (as well as their numerous varieties). In petrology the term “chert” is used to generally refer to all rocks composed primarily of microcrystalline, cryptocrystalline and microfibrous quartz. The term does not include quartzite. Chalcedony is a microfibrous (microcrystaline with a fibrous structure) variety of quartz. Strictly speaking, the term “flint” is reserved for varieties of chert which occur in chalk and marly limestone formations. Among non-geologists (in particular among archaeologists), the distinction between "flint" and "chert" is often one of quality—chert being lower quality than flint. This usage of the terminology is prevalent in America and is likely caused by early immigrants who imported the terms from England where most true flint (that found in chalk formations) was indeed of better quality than "common chert" (from limestone formations). Among petrologists, chalcedony is sometimes considered separately from chert due to its fibrous structure. Since many cherts contain both microcrystaline and microfibrous quartz, it is sometimes difficult to classify a rock as completely chalcedony, thus its general inclusion as a variety of chert.

Chert and Precambrian Fossils

The cryptocrystalline nature of chert, combined with its above average ability to resist weathering, recrystallisation and metamorphism has made it an ideal rock for preservation of early life forms.

For example:

- The 3.2 billion year old chert of the figure Tree Formation in the Barbeton Mountains between Swaziland and South Africa preserved non-colonial unicellular bacterial-like fossils.
- The Gunflint Chert of western Ontario (1.9 to 2.3 BYA) preserves not only bacteria and cyanobacteria but also organisms believed to be ammonia-consuming and some that resemble green algae and fungus-like organisms.
- The Apex Chert (3.4 BYA) of the Pilbara craton, Australia preserved eleven taxa of prokaryotes.
- The Bitter Springs Formation of the Amadeus Basin, Central Australia, preserves 850 MYA old cyanobacteria and algae.
- The Devonian Rhynie chert (400 MYA) of Scotland has the oldest remains of land flora, and the preservation is so perfect that it allows cellular studies of the fossils.

In prehistoric times, chert was often used as a raw material for the construction of stone tools. Like obsidian, as well as some rhyolites, felsites, quartzites, and other tool stones used in lithic reduction, chert fractures in a Hertzian cone when struck with sufficient force. This results in conchoidal fractures, a characteristic of all minerals with no cleavage planes. In this kind of fracture, a cone of force propagates through the material from the point of impact, eventually removing a full or partial cone; this result is familiar to anyone who has seen what happens to a plate-glass window when struck by a small object, such as an airgun projectile. The partial Hertzian cones produced during lithic reduction are called flakes, and exhibit features characteristic of this sort of breakage, including striking platforms, bulbs of force, and occasionally eraillures, which are small secondary flakes detached from the flake's bulb of force.

When a chert stone is struck against steel, sparks result. This makes it an excellent tool for starting fires, and both flint and common chert were used in various types of fire-starting tools, such as tinderboxes, throughout history. A primary historic use of common chert and flint was for flintlock firearms, in which the chert striking a metal plate produces a spark that ignites a small reservoir containing black powder, discharging the firearm.

In some areas, chert is ubiquitous as stream gravel and fieldstone and is currently used as construction material and road surfacing. Part of chert's popularity in road surfacing or driveway construction is that rain tends to firm and compact chert while other fill often gets muddy when wet. However, where cherty gravel ends up as fill in concrete, the slick surface can cause localized failure. Hauled prices for chert of less than $10 to $15 per ton are not uncommon in many parts of the U.S.

Chert has been used in late nineteenth-century and early twentieth-century headstones or grave markers in Tennessee and other regions.

Varieties of Chert

There are numerous varieties of chert, classified based on their visible, microscopic and physical characteristics. Some of the more common varieties are:

- Flint is compact microcrystalline quartz. It is found in chalk or marly limestone formations and is formed by a replacement of calcium carbonate with silica. It is commonly found as nodules. This variety was often used in past times to make bladed tools.
- "Common chert" is a variety of chert which forms in limestone formations by replacement of calcium carbonate with silica. This is the most abundantly found variety of chert. It is generally considered to be less attractive for producing gem stones and bladed tools than flint.
- Jasper is a variety of chert formed as primary deposits, found in or in connection with magmatic formations which owes its red color to iron(III) inclusions. Jasper frequently also occurs in black, yellow or even green (depending on the type of iron it contains). Jasper is usually opaque to near opaque.
- Radiolarite is a variety of chert formed as primary deposits and containing radiolarian microfossils.
- Chalcedony is microfibrous quartz.
- Agate is distinctly banded chalcedony with successive layers differing in color or value.
- Onyx is a banded agate with layers in parallel lines, often black and white.
- Opal is a hydrated silicon dioxide. It is often of a Neogenic origin. In fact is not a mineral (it is a mineraloid) and it is generally not considered a variety of chert,

although some varieties of opal (opal-C and opal-CT) are microcrystalline and contain much less water (sometime none). Often people without petrological training confuse opal with chert due to similar visible and physical characteristics.

- Magadi-type chert is a variety that forms from a sodium silicate precursor in highly alkaline lakes such as Lake Magadi in Kenya.

Other lesser used terms for chert (most of them archaic) include, firestone, silex, silica stone and flintstone.

Flint

Flint (or flintstone) is a hard, sedimentary, cryptocrystalline form of the mineral quartz, categorized as a variety of chert. It occurs chiefly as nodules and masses in sedimentary rocks, such as chalks and limestones. Inside the nodule, flint is usually dark gray, black, green, white, or brown in color, and often has a glassy or waxy appearance. A thin layer on the outside of the nodules is usually different in color, typically white, and rough in texture. From a petrological point of view, "flint" refers specifically to the form of chert that occurs in chalk or marly limestone. Similarly, "common chert" (sometimes referred to simply as "chert") occurs in limestone.

A flint nodule from the Onondaga limestone layer, Buffalo, New York. (3.8 cm wide).

In early history, flint was used for making cutting tools and igniting fire. In addition, it has been used as a material for the construction of houses, churches, and other structures. It has also been used in the preparation of ceramics.

Formation

Flint is a form of quartz, found as masses associated with chalk and limestone. However, the exact mode of formation of flint is not yet clear. It is currently thought that flint is formed as a result of chemical changes in compressed sedimentary rock formations, during the process of diagenesis. One hypothesis is that a gelatinous material fills cavities in the sediment, such as holes bored by crustaceans or molluscs and that this becomes silicified. This theory is able to explain the complex shapes of flint nodules that are found. The source of dissolved silica in the porous media could arise from the spicules of siliceous sponges.

Pebble beach made up of flint nodules eroded out of the nearby chalk cliffs, Cape Arkona, Rügen.

Late Stone Age flint axe, about 31 cm long.

Uses

Tools or Cutting Edges

Flint was used for the manufacture of flint tools during the Stone Age, as it splits into thin, sharp splinters called flakes or blades (depending on the shape) when struck by another hard object (such as a hammerstone made of another material). This process is referred to as knapping.

Modern flint spark lighter being struck.

In Europe, some of the best toolmaking flint has come from Belgium (Obourg, flint mines of Spiennes), the coastal chalks of the English Channel, the Paris Basin, Thy in Jutland (flint mine at Hov), the Sennonian deposits of Rügen, Grimes Graves in England and the Jurassic deposits of the Kraków area in Poland. Flint mining is attested since the Paleolithic, but became more common since the Neolithic (Michelsberg culture, Funnelbeaker culture).

Detail of flint used in a building in Wiltshire, England.

To Ignite Fire or Gunpowder

When struck against steel, a flint edge will produce sparks of steel. The hard flint edge shaves off a particle of the steel that, heated by the friction, burns with oxygen from the atmosphere and can ignite the proper tinder. This method is popular in woodcraft, bushcraft, and among others who wish to use traditional skills.

Striking sparks with flint and steel is not particularly easy or convenient method to start a fire, although it is much easier than other primitive fire-making methods such as using a bow drill. As with most skills, practice improves results. By comparison, matches and cigarette lighters are a much quicker and more convenient way of starting a fire.

A later, major use of flint and steel was the flintlock, primarily used on firearms. A piece of flint held in the jaws of a spring-loaded hammer, when released by a trigger, strikes a hinged piece of steel ("frizzen") at an angle, creating a shower of sparks and exposing a charge of priming powder. The sparks ignite the priming powder and that flame, in turn, ignites the main charge propelling the ball, bullet, or shot in the barrel. While the military use of the flintlock declined after the adoption of the percussion cap from the 1840s onward, the flintlock is still popular on hunting rifles and shotguns used in the United States.

Use of flint and steel should not be confused with use of ferrocerium. This human-made material, when scraped with any hard, sharp edge, produces sparks that are much hotter than obtained with natural flint and steel, allowing use of a wider range of tinder. Because it can produce sparks when wet and can start hundreds or thousands of fires when used correctly, ferrocerium is a common item included in survival kits. Called "flint," ferrocerium is also used in many cigarette lighters.

As a Building Material

Flint, knapped or unknapped, has been used since antiquity (for example at the Late Roman fort of Burgh Castle in Norfolk) up to the present day as a material for building stone walls, using lime mortar, and often combined with other available stone or brick rubble. It was most common in parts of southern England, where no good building stone was available locally, and brick-making not widespread until the later Middle Ages. It is especially associated with East Anglia, but also used in chalky areas stretching through Sussex, Surrey and Kent to Somerset. Flint was used in the construction of many churches, houses, and other buildings, for example the large stronghold of Framlingham Castle. Many different decorative effects have been achieved by using different types of knapping or arrangement and combinations with stone (flushwork), especially in the fifteenth and early sixteenth centuries.

A flint church—the Parish Church of Saint Thomas, in Cricket Saint Thomas, Somerset, England. The height of the very neatly knapped flints varies between 3 and 5 inches (8-13 cm).

Ceramics

Flint pebbles are used as the media in ball mills to grind glazes and other raw materials for the ceramics industry. The pebbles are hand-selected for color, with those showing a reddish tint, indicating the presence of iron, being discarded. The remaining blue-gray stones have a low content of chromophoric oxides and so should impart lesser amounts of coloring contaminants.

In the UK, flint pebbles were traditionally an important raw material for clay-based ceramic bodies. After calcination to remove organic impurities and induce certain physical reactions, and milling to fine particle size, flint was added as filler to pottery bodies. However, flint is no longer used and has been replaced by quartz as is used in other countries. Because of this historical use, the word “flint” is used by U.S. potters to refer to siliceous materials that are not flint.

Dolomites

Dolomite is a common rock-forming mineral. It is a calcium magnesium carbonate with a chemical composition of $CaMg(CO_3)_2$. It is the primary component of the sedimentary rock known as dolostone and the metamorphic rock known as dolomitic marble. Limestone that contains some dolomite is known as dolomitic limestone.

Dolomite is rarely found in modern sedimentary environments, but dolostones are very common in the rock record. They can be geographically extensive and hundreds to thousands of feet thick. Most rocks that are rich in dolomite were originally deposited as calcium carbonate muds that were postdepositionally altered by magnesium-rich pore water to form dolomite.

Dolomite crystals: Dolomite crystals from Penfield, New York. This specimen is approximately 3 inches (6.7 centimeters) across.

Dolomite is also a common mineral in hydrothermal veins. There it is often associated with barite, fluorite, pyrite, chalcopyrite, galena, or sphalerite. In these veins it often occurs as rhombohedral crystals which sometimes have curved faces.

Granular Dolomite: Dolomitic marble from Thornwood, New York. This specimen is approximately 3 inches (6.7 centimeters) across.

Physical Properties of Dolomite

The physical properties of dolomite that are useful for identification are presented. Dolomite has three directions of perfect cleavage. This may not be evident when the dolomite is fine-grained. However, when it is coarsely crystalline the cleavage angles can easily be observed with a hand lens. Dolomite has a Mohs hardness of 3 1/2 to 4 and is sometimes found in rhombohedral crystals with curved faces. Dolomite produces a very weak reaction to cold, dilute hydrochloric acid; however, if the acid is warm or if the dolomite is powdered, a much stronger acid reaction will be observed. (Powdered dolomite can easily be produced by scratching it on a streak plate.)

Dolomite is very similar to the mineral calcite. Calcite is composed of calcium carbonate ($CaCO_3$), while dolomite is a calcium magnesium carbonate ($CaMg(CO_3)_2$). These two minerals are one of the most common pairs to present a mineral identification challenge in the field or classroom.

Dolostone: Dolostone from Lee, Massachusetts. The "sugary" sparkle displayed by this rock is caused by light reflecting from tiny dolomite cleavage faces. This specimen is approximately 4 inches (10 centimeters) across.

The best way to tell these minerals apart is to consider their hardness and acid reaction. Calcite has a hardness of 3, while dolomite is slightly harder at 3 1/2 to 4. Calcite is also strongly reactive with cold hydrochloric acid, while dolomite will effervesce weakly with cold hydrochloric acid.

Physical Properties of Dolomite	
Chemical Classification	Carbonate
Color	Colorless, white, pink, green, gray, brown, black.
Streak	White
Luster	Vitreous, pearly
Diaphaneity	Transparent to translucent
Cleavage	Perfect, rhombohedral, three directions.
Mohs Hardness	3.5 to 4
Specific Gravity	2.8 to 2.9
Diagnostic Properties	Rhombohedral cleavage, powdered form effervesces weakly in dilute HCl, hardness.
Chemical Composition	$CaMg(CO_3)_2$
Crystal System	Hexagonal
Uses	Construction aggregate, cement manufacture, dimension stone, calcined to produce lime, sometimes an oil and gas reservoir, a source of magnesia for the chemical industry, agricultural soil treatments, metallurgical flux.

Solid Solution and Substitution

Dolomite occurs in a solid solution series with ankerite ($CaFe(CO_3)_2$). When small amounts of iron are present, the dolomite has a yellowish to brownish color. Dolomite and ankerite are isostructural.

Kutnahorite ($CaMn(CO_3)_2$) also occurs in solid solution with dolomite. When small amounts of manganese are present, the dolomite will be colored in shades of pink. Kutnahorite and dolomite are isostructural.

Dolomite aggregate: These specimens are approximately 1/2 inch to 1 inch (1.3 centimeters to 2.5 centimeters) across.

Uses of Dolomite

Dolomite as a mineral has very few uses. However, dolostone has an enormous number of uses because it occurs in deposits that are large enough to mine.

This specimen is approximately 4 inches (10 centimeters) cross.

The most common use for dolostone is in the construction industry. It is crushed and sized for use as a road base material, an aggregate in concrete and asphalt, railroad ballast, rip-rap, or fill. It is also calcined in the production of cement and cut into blocks of specific size known as "dimension stone."

Dolomite's reaction with acid also makes it useful. It is used for acid neutralization in the chemical industry, in stream restoration projects, and as a soil conditioner.

Dolomite is used as a source of magnesia (MgO), a feed additive for livestock, a sintering agent and flux in metal processing, and as an ingredient in the production of glass, bricks, and ceramics.

Dolomite serves as the host rock for many lead, zinc, and copper deposits. These deposits form when hot, acidic hydrothermal solutions move upward from depth through a fracture system that encounters a dolomitic rock unit. These solutions react with the dolomite, which causes a drop in pH that triggers the precipitation of metals from solution.

Dolomite also serves as an oil and gas reservoir rock. During the conversion of calcite to dolomite, a volume reduction occurs. This can produce pore spaces in the rock that can be filled with oil or natural gas that migrate in as they are released from other rock units. This makes the dolomite a reservoir rock and a target of oil and gas drilling.

Coal

Coal is an organic sedimentary rock that forms from the accumulation and preservation of plant materials, usually in a swamp environment. Coal is a combustible rock and, along with oil and natural gas, it is one of the three most important fossil fuels. Coal has a wide range of uses; the most important use is for the generation of electricity.

Bituminous Coal.

In figure, Bituminous coal is typically a banded sedimentary rock. In this photo you can see bright and dull bands of coal material oriented horizontally across the specimen. The bright bands are well-preserved woody material, such as branches or stems. The dull bands can contain mineral material washed into the swamp by streams; charcoal produced by fires in the swamp, or degraded plant materials. This specimen is approximately three inches across (7.5 centimeters).

Formation of Coal

Coal forms from the accumulation of plant debris, usually in a swamp environment. When a plant dies and falls into the swamp, the standing water of the swamp protects it from decay. Swamp waters are usually deficient in oxygen, which would react with the plant debris and cause it to decay. This lack of oxygen allows the plant debris to persist. In addition, insects and other organisms that might consume the plant debris on land do not survive well under water in an oxygen-deficient environment.

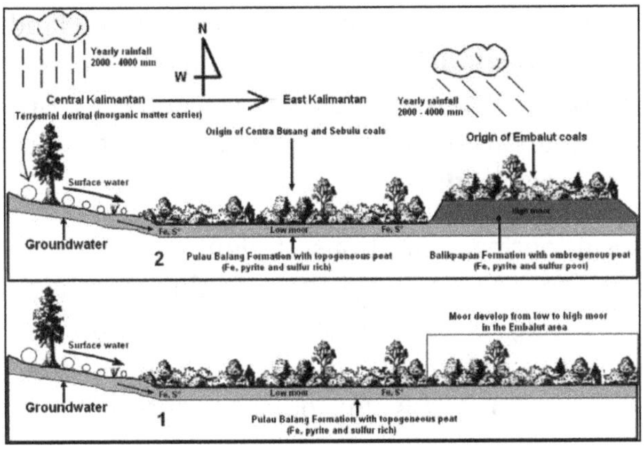

Coal-Forming Environments.

To form the thick layer of plant debris required to produce a coal seam, the rate of plant debris accumulation must be greater than the rate of decay. Once a thick layer of plant debris is formed, it must be buried by sediments such as mud or sand. These are typically washed into the swamp by a flooding river. The weight of these materials compacts the plant debris and aids in its transformation into coal. About ten feet of plant debris will compact into just one foot of coal.

Plant debris accumulates very slowly. So, accumulating ten feet of plant debris will take a long time. The fifty feet of plant debris needed to make a five-foot thick coal seam would require thousands of years to accumulate. During that long time, the water level of the swamp must remain stable. If the water becomes too deep, the plants of the swamp will drown, and if the water cover is not maintained the plant debris will decay. To form a coal seam, the ideal conditions of perfect water depth must be maintained for a very long time.

Peat: A mass of recently accumulated to partially carbonized plant debris. This material is on its way to becoming coal, but its plant debris source is still easily recognizable.

If you are an astute reader you are probably wondering: "How can fifty feet of plant debris accumulate in water that is only a few feet deep?" The answer to that question is the primary reason that the formation of a coal seam is a highly unusual occurrence. It can only occur under one of two conditions: 1) a rising water level that perfectly keeps pace with the rate of plant debris accumulation; or, 2) a subsiding landscape that perfectly keeps pace with the rate of plant debris accumulation. Most coal seams are thought to have formed under condition #2 in a delta environment. On a delta, large amounts of river sediments are being deposited on a small area of Earth's crust, and the weight of those sediments causes the subsidence.

For a coal seam to form, perfect conditions of plant debris accumulation and perfect conditions of subsidence must occur on a landscape that maintains this perfect balance for a very long time. It is easy to understand why the conditions for forming coal have occurred only a small number of times throughout Earth's history. The formation of a coal requires the coincidence of highly improbable events.

Coal "Rank"

Plant debris is a fragile material compared to the mineral materials that make up other rocks. As plant debris is exposed to the heat and pressure of burial, it changes in composition and properties. The "rank" of a coal is a measure of how much change has occurred. Sometimes the term "organic metamorphism" is used for this change.

Based upon composition and properties, coals are assigned to a rank progression that corresponds to their level of organic metamorphism.

Lignite: The lowest rank of coal is "lignite." It is peat that has been compressed, dewatered, and lithified into a rock. It often contains recognizable plant structures.

Uses of Coal

Coal-Fired Power Plant.

Electricity production is the primary use of coal in the United States. Most of the coal mined in the United States is transported to a power plant, crushed to a very small particle size, and burned. Heat from the burning coal is used to produce steam, which turns a generator to produce electricity. Most of the electricity consumed in the United States is made by burning coal.

In figure, power plant where coal is burned to produce electricity. The three large stacks are cooling towers where water used in the electricity generation process is cooled before reuse or release to the environment. The emission streaming from the right-most stack is water vapor. The combustion products from burning the coal are released into the tall, thin stack on the right. Within that stack are a variety of chemical sorbents to absorb polluting gases produced during the combustion process.

Coal has many other uses. It is used as a source of heat for manufacturing processes. For example, bricks and cement are produced in kilns heated by the combustion of a jet of powdered coal. Coal is also used as a power source for factories. There it is used to heat steam, and the steam is used to drive mechanical devices. A few decades ago most coal was used for space heating. Some coal is still used that way, but other fuels and coal-produced electricity are now used instead.

Coke production remains an important use of coal. Coke is produced by heating coal under controlled conditions in the absence of air. This drives off some of the volatile materials and concentrates the carbon content. Coke is then used as a high-carbon fuel for metal processing and other uses where an especially hot-burning flame is needed.

Anthracite coal: Anthracite is the highest rank of coal. It has a bright luster and breaks with a semi-conchoidal fracture.

Coal is also used in manufacturing. If coal is heated the gases, tars, and residues produced can be used in a number of manufacturing processes. Plastics, roofing, linoleum, synthetic rubber, insecticides, paint products, medicines, solvents, and synthetic fibers all include some coal-derived compounds. Coal can also be converted into liquid and gaseous fuels; however, these uses of coal are mainly experimental and done on a small scale.

Limestones

Limestone is a very common sedimentary rock consisting of calcium carbonate (more than 50%). It is the most common non-siliciclastic (sandstone and shale are common siliciclastic rocks) sedimentary rock. Limestones are rocks that are composed of mostly calcium carbonate (minerals calcite or aragonite).Carbonate rocks where the dominant carbonate is dolomite (calcium magnesium carbonate) are named dolomite rock. Carbonate rocks together make up one-fifth of all sedimentary rocks in the stratigraphic record.

Varieties of limestone.

In figure, Chalk (upper left) is a marine limestone consisting of tests of microscopic algae and foraminifera. Tufa (upper right) is a chemical precipitate of calcium carbonate. Fossils are very common in marine calcitic sedimentary rocks. Rocks such as coquina are wholly composed of fossils but so-called normal limestones may be also highly fossiliferous. The sample (lower left) is from the Ordovician. Grainstone is a coarse-grained grain-supported variety that contains almost no limy mud (micrite).

Limestone is defined by these two criteria: it is a sedimentary rock and it is composed of calcium carbonate. There are other rocks that are composed of calcium carbonate. Carbonatite is a rare type of igneous rock and marble is a common metamorphic rock. Both are chemically composed of calcium carbonate, but they are not limestones because they are not sedimentary rocks.

Limestone formations are usually clearly layered. Pure rock is light-colored.
An outcrop of Silurian limestone in Saaremaa, Estonia.

Other than that, there are no more restrictions. Hence, limestones can be, and indeed are, formed by very different mechanisms. Some limestones are clastic (or detrital) just like sandstone. However, there is an important difference. Most detrital limestones are made of biogenic grains rather than weathered bits and pieces of preexisting rocks. Calcium carbonate in the form of aragonite and calcite are extremely common biominerals. They are used to build tests, shells, exoskeletons, etc. of millions of marine (mostly) species. These tests sink to the bottom of oceans or other waterbodies and form limy sediment (named ooze if the tests are microscopic) which becomes limestone after burial and lithification. However, some detrital limestones (oolites) are composed of abiogenic particles like ooids. Biogenic limestone may grow in situ by the growth of carbonate skeletons (coral reefs) or by trapping of sediments in bacterial mats (stromatolites).

Stevns Klint in Denmark is a world-famous K-T boundary.

In figure, it is composed of bryozoan limestone (upper part) and chalk (lower) which are both types of limestone. These layers are separated by a narrow layer of clayish material which formed 65 million years ago. The clay is very rich in iridium – a metal abundant in iron meteorites but rare on Earth. This layer formed as a result of meteorite explosion on the Yucatán Peninsula in Mexico. The event marks the end of the Mesozoic Era and is associated with the mass extinction that also ended the reign of dinosaurs. Dark gray spots are chert nodules and the boulders on the coast are also made of this rock type.

Limy mud is often purely chemical precipitate. Tufa and travertine are varieties of chemical limestones. Tufa is soft and porous variety that is associated with springs.

Ground water may contain thousand times as much carbon dioxide as would be in equilibrium with air. Such groundwater is acidic and dissolves limestone. Limestone (or more precisely limy mud) precipitates out of ground water when it emerges above ground as a spring and loses much of its CO_2. Travertine is a hard variety of chemical limestone. Travertine is the material speleothems (stalactites, stalagmites, and other dripstones) of karst caves are made of. Travertine formed in these caves is simply the result of a complete evaporation of water containing mainly calcium carbonate. Travertine deposited from hot springs may be a result of the loss of carbon dioxide from the water as pressure is released upon emerging at the surface.

Limestones are strongly affected by diagenetic processes that accompany lithification. Aragonite will be replaced by calcite (if the original carbonate mineral was aragonite), dissolution may be significant (there is a picture of stylolite below), and calcite may be replaced with dolomite. However, calcite is resistant to metamorphism. Marble is a metamorphosed limestone, but it is still composed of carbonate minerals. The crystals are just bigger, they are recrystallized, and original structures (like fossils) are obliterated.

Limestone is often impure. Calcium carbonate may be mixed with siliciclastic material (clay, sand) in all proportions, but calcium carbonate must form the majority. Otherwise, we have calcareous sandstone or marl instead of limestone. The term "marl" is mostly used in the field and often somewhat muddy limestones, even if they contain more calcium carbonate than mud, are called that way.

Limestone is often rich in fossils. Sea lilies (attached crinoids) are echinoderms that are not extinct, but their best days have passed. They were so abundant during the Paleozoic that many limestones are full of their fossils. Bits of crinoids in a limestone slab from the Ordovician of Estonia.

Limestone has many industrial uses. It is used as a building material, as a raw material in the manufacture of portland cement, quicklime, etc. Calcium is an important micronutrient. Several commercial drugs (antacids and calcium supplements) are nothing but pulverized limestone (calcite). It is an important reservoir rock of crude oil and gas.

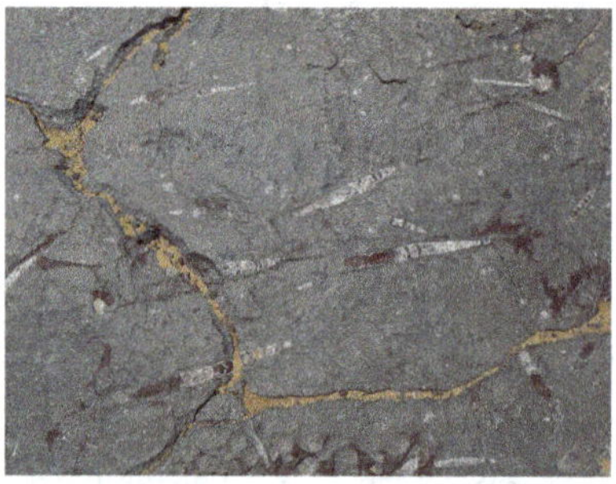

Limestone from the Ordovician of Morocco containing lots of nautiloid fossils.

Mudrocks

Mudrocks are fine grained sedimentary rocks consisting of mostly silt and clay size fragments. They are sometimes called argillites. Because of their small grain size, they are difficult to study, even with the petrographic microscope. But, they are important rocks because they are the most abundant sedimentary rocks, making up over 65% all sedimentary rocks, are likely the source rocks for petroleum and natural gas, and are sometimes valuable ore deposits. In addition, the mudrocks are the protoliths (precursor rock) for aluminous metamorphic rocks, often called pelitic metamorphic rocks.

Classification

Classification of mudrocks is mainly based on observations one can make in the field or at the level of a hand specimen. The classification depends on the grain size of the minerals making up the rock and whether or not the rock is fissile or non-fissile. A fissile rock tends to break along sheet-like planes that are nearly parallel to the bedding planes. Fissility is caused by the tendency of clay minerals to be deposited with their sheet structures [(001) crystallographic planes] parallel to the depositional surface.

Grain Size	Description	Fissile Rock	Nonfissile Rock
>2/3 silt	Abundant silt sized grains visible with a hand lens	Silt-shale	Siltstone
>1/3, <2/3 silt	Feels gritty when chewed	Mud-shale	Mudstone
>2/3 clay	Feels smooth when chewed	Clay-shale	Claystone

Texture

Elements of texture that can be observed in mudrocks include the shapes of the grains, the fissility or lack of fissility, and laminations.

- Grain shape: Clay minerals are generally angular and sheet like, reflecting their crystal structure, which is also why most mudrocks are fissile. Silt size quartz grains are usually angular or platy shaped. Little rounding occurs because these minerals are so small that they can be kept in suspension in very low energy currents, with little chance of impact with other grains that would normally cause abrasion and rounding.
- Fissility: Whether or not a mudrock is fissile or non-fissile depends on several factors.
  - The abundance of clay minerals. The more clay minerals contained in the rock, the more likely the rock is to be fissile.
  - The degree of preferred orientation of the clay minerals. Small clay minerals tend to adhere to one another if they collide during transport. The tendency is increased by increased salinity of the water and the presence of organic matter in the water. Adhesion of small mineral grains is referred to as *flocculation*. If the clay minerals flocculate, then they are less likely to have a preferred orientation, and thus less likely to form rocks with fissility.

    - Bioturbation of organisms within or on the surface of the sediment can disturb the preferred orientation of clay minerals, and thus lead to a non-fissile rock.
    - If the clay minerals recrystallize during diagenesis they will tend to do so with a preferred orientation with their (001) crystal planes oriented perpendicular to the maximum principal stress direction. (This process also results in slatey cleavage and foliation in metamorphic rocks). If diagenesis occurs shortly after deposition, then it is likely that the maximum principal stress direction will also be oriented perpendicular to the bedding planes.
- Laminations: Laminations are parallel layers less than 1 cm thick. Such laminations can be seen as differences in grain size of the clasts in different laminae - due to changes in current velocity of the depositing medium, or could be due to changes in the organic content and oxidation conditions at the site of deposition of the different layers.

Color

Most mudstones fall into two classes in terms of their color:

- Gray-Black: Gray to black color of mudrocks usually indicates the presence of more than 1% organic matter in the form of carbon or some carbon compound. If organic matter occurs in the sediment and there is an abundance of oxygen, the carbon would rapidly be oxidized to CO_2. Thus the presence of reduced carbon indicates a reducing condition at the time of deposition or burial. Such reducing conditions usually occur in stagnant water where little circulation takes place. This can occur in restricted seas, in swamps, and in some lakes and lagoons. Reducing conditions are not usually present in the deep oceans, because of the thermohaline circulation of ocean waters that brings water that originated at the surface in the Polar Regions to the bottoms of the ocean basins. Rapid burial of sediment can also lead to reducing conditions and preservation of carbon and carbon compounds in the rock.
- Red-Brown-Yellow-Green: Red, brown, yellow, and green colorations are a reflection of the oxidation state of Fe in the sediments. Under oxidizing conditions most of the Fe will be Fe^{+3} and give the sediment a red coloration as in Hematite (Fe_2O_3). If the Fe^{+3}occurs as Iron hydroxide [FeO(OH)] (geothite) it gives the sediment a brown coloration. Limonite gives the sediment a yellow color. The absence of oxidized Fe allows the green color given to minerals containing Fe^{+2} to show through. Thus, red, brown, and yellow colored mudrocks reflect deposition in an oxidizing environment. Note, however, that it requires very little hematite in a rock to give the rock a red stain or coloration.

Mineralogy

Clay minerals are the most abundant minerals in mudstones, making up over 60% of all mudstones. Other minerals like quartz, feldspar, carbonate minerals, organic compounds (not really minerals), sulfides, and hematite also occur.

- Clay Minerals: Three groups of clay minerals occur in sedimentary rocks, and are most easily distinguished using x-ray diffraction techniques as discussed in your mineralogy class.
    - Kaolinites [$Al_2Si_2O_5(OH)_4$] are formed in warm moist climates where Ca, Na, and K ions are leached and removed in solution during the weathering process. Thus, the presence of kaolinite clays indicates a source in a humid tropical climate.
    - Smectites are expanding clays. That is they can expand by taking in water between layers. Montmorillinite-$(\frac{1}{2}Ca,Na)_{0.7}(Al,Fe,Mg)_4Si,Al)_8O_{20}(OH)_4 \cdot nH_2O$ is a good example. These clays form from weathering of Fe -Mg rich igneous and metamorphic rocks in temperate climates, and are the most abundant clays in modern sediment.
    - Illites- $K1_{-1.5}Al_4Si_{7-6.5}Al_{1-1.5}O_{20}(OH)_4$ - These clays are formed by weathering of feldspars in temperate climates and by alteration of smectite clays during diagenesis. They have a structure similar to muscovite.
    - Mixed layer clays are clays that show interlayering between smectites like layers and illite like layers in the same crystal. These are also common clays in the modern sediment.

If one looks at mudrocks of various ages, it is observed that Tertiary and younger mudrocks consist mostly of smectites and mixed layer clays, In older mudrocks illite is the dominant clay mineral, and in early Paleozoic and older rocks less than 10% of the clay minerals are smectite clays. This is likely due to diagenetic processes that convert the smectites to illites over time. Such a conversion would involve the release of elements like Si, Ca, Na, Mg, and Fe from the clays. Most of these released elements likely remain in the mudrock and crystallize to form quartz, chlorite, albite, calcite, dolomite, siderite, and ankerite, which are all minerals that occur in mudrocks. Many of these minerals may in fact be products of diagenesis.

- Quartz: Most of the quartz that occurs in mudrocks is single crystals. Most of the silt-sized fraction is quartz, but very little of the clay sized fraction is quartz. Single crystal of quartz may originate as detrital grains produced by abrasion of larger grains, may be diagenetically produced during the smectite to illite transformation, or may be recrystallized remains of silica secreting organisms like diatoms and radiolaria.

- Feldspar: Feldspars also occur in the silt-sized fraction of most mudstones, but in very small amounts. The lack of feldspars in mudrocks in comparison to river sediment is due to the unstable nature of feldspar - it continues to weather even after deposition, and due to diagenetic changes that take place in the sediment.
- Carbonate Minerals: Carbonate minerals occur in mudrocks with about the same abundance as feldspars. Most is probably calcite and is produced either by diagenesis or as detrital grains produced by foraminefera or other microorganisms.
- Organic Matter: As discussed above, deposition in waters with restricted circulation or rapid burial can result in reducing conditions and the preservation of carbon and carbon compounds in mudrocks. In some cases, like oil shales, the carbon compounds undergo further transformations that convert them to hydrocarbons like petroleum. Otherwise, the carbon or carbon compounds generally tend to give the rock a dark gray to black color.
- Others: Other minerals, like sulfide minerals, including pyrite, are found in black shales, again produced under reducing conditions, and hematite, likely indicating oxidizing conditions during deposition or diagenesis.
- Bentonites: One particular type of mudrock consists of mostly of white colored smectite clays and colloidal silica. These are produced by the alteration of siliceous fine-grained volcanic ash. These often contain small amounts of quartz and sanidine, which likely represented phenocrysts in the ash producing magma. Bentonite layers usually represent pyroclastic fall deposits, and as such they make excellent marker beds, both because they are widespread, and because the sanidine in the ash can is dated by K-Ar methods.

Environments of Deposition

Mudrocks represent texturally and mineralogically mature sediments deposited in low energy environments. The fine grain size of these sediments means that that the sediment can be suspended for long times in relatively quiet, low energy currents. This results in deposition on the abyssal plains of the oceans, at the distal ends of deltas, in quiet lakes and swamps, and as wind-blown dust. Large deposits of wind-blown dust are called loess, and consist of mostly silt-sized fragments. While the small fragments can be transported into lakes and oceans easily by streams, it is likely that large quantities of these fragments could also reach the oceans by wind transport.

Carbonate Rocks

Carbonate rocks are made of particles (composed >50% carbonate minerals) embedded in a cement. Most carbonate rocks result from the accumulation of bioclasts created by calcareous organisms. Therefore carbonate rocks originate in area favoring biological activity i.e. in shallow and warm seas in areas with little to no siliciclastic input. In present day Earth these areas are limited to ±40 latitude in region away or protected from erosion-prone elevated continental areas.

In the Devonian high-sea level, favoring the development of shallow epicontinental sea, warmer conditions, and relatively low detrital sedimentation rates, resulted in larger thick carbonate buildups. The Murrumbidgee serie is an example of such carbonate buildup.

Because they contain numerous bioclasts (i.e. fossils) carbonate rocks are cherish by stratigraphers and paleontologist alike as they enable them to reconstruct stratigraphic sequences and make large-scale correlation between geological formations hundreds to thousands of kilometer apart. In addition carbonate rocks' texture and the nature of their bioclasts offer a detailed insight into their depositional environment.

Depositional Environments

Carbonate depositional environment past or present fall into three general types:

- Ramp continental margins are continental platform gently sloping toward the ocean (<1). They are limited by the emerged continent and toward the oceanic basin by the gentle continental break. It is an environment of high energy as the amplitude of the waves increases as the depth of the sea decreases.
- Rimmed margins are continental platform limited toward the ocean by a steep and abrupt continental break where a nearly continuous carbonate rim or barrier develops. These rims or barriers are wave resistant structure made of reef coral (alive or dead) and oolitic sand shoals. Landward the rim/barrier is a low energy environment of "lagoonal" characteristics grading landward into a tidal flat. In present Earth they can be recognized by colorful tourists snorkeling happily together. Oceanward, the rim/barrier is a high-energy environment. In this noisy, wavy and scummy environment it takes commitment, strength and a bit of an attitude to any living organism to hold on onto the place. Tourists are rarely seen in those dangerous waters.
- Isolated platforms, also known as "Bahama type", are shallow platform 10's to 100's km wide offshore of shallow continental shelves, surrounded by 100's to 1000's m deep water. This environment is characterized by the absence of siliciclastic input.

Dunham Classification

The Dunham classification is based on concept of grain/mud support therefore on the proportion mud-particle and the depositional textures. The concept of "support" assumes continuity of either the mud matrix or that of the grains. If the carbonate is mud-supported the grains float into a continuum of mud matrix. In grain-supported carbonate rocks the grains from an interconnected skeleton in which the mud fills the gap. The percentage mud/cement at which there is a switch between mud-supported and grain-supported depends on the fabric (preferential agencement) of the particles.

Deposition Texture Recognizable					Unrecognizable
Original components not bound together during deposition				Original Bound	
Contain mud, clay, and fine silt-size carbonate			No Mud		
Mud Supported		Grain-supprtd			
Grains < 10%	Grains > 10%				
Mudstones	Wakestone	Packstone	Grainstone	Boundstone	Crystalline

Mudstone

A mud-supported carbonate sedimentary rock containing less than 10% particles of clay and fine silt size; the original components are not bound together during deposition.

Wakestone

Mud-supported limestone containing more than 10% carbonate grains (larger than 20 microns) “floating” in a finer-grained lime mud matrix.

References

- Sedimentary-rock, science: britannica.com, Retrieved 1 April, 2019
- Sedimentary-environment, almanacs-transcripts-and-maps, science: encyclopedia.com, Retrieved 12 March, 2019
- Breccia-rock: thoughtco.com, Retrieved 6 January, 2019
- Conglomerate: geologylearn.blogspot.com, Retrieved 16 April, 2019
- Siltstone, rocks: geology.com, Retrieved 23 June, 2019
- Shale, rocks: geology.com, Retrieved 3 February, 2019
- Sandstone, minerals-database: mineralseducationcoalition.org, Retrieved 13 July, 2019
- Iron-ore, rocks: geology.com, Retrieved 7 May, 2019
- Chert: newworldencyclopedia.org, Retrieved 18 March, 2019
- Flint: newworldencyclopedia.org, Retrieved 30 June, 2019
- Dolomite, minerals: geology.com, Retrieved 9 February, 2019
- Coal, rocks: geology.com, Retrieved 19 July, 2019
- Limestone: sandatlas.org, Retrieved 2 May, 2019
- Mudrocks: tulane.edu, Retrieved 3 August, 2019
- Carbonates: geosci.usyd.edu.au, Retrieved 13 January, 2019

4

Metamorphic Rocks

The rocks which arise from the transformation of existing rock types through a process called metamorphism are known as metamorphic rocks. Some of the different types of metamorphism are contact or thermal metamorphism, regional metamorphism, hydrothermal metamorphism and plutonic metamorphism. The chapter closely examines these key types of metamorphic rocks to provide an extensive understanding of the subject.

Metamorphic rock are any of a class of rocks that result from the alteration of preexisting rocks in response to changing environmental conditions, such as variations in temperature, pressure, and mechanical stress, and the addition or subtraction of chemical components. The preexisting rocks may be igneous, sedimentary, or other metamorphic rocks.

Slate: Slate is a fine-grained clayey metamorphic rock that cleaves, or splits, readily into thin slabs having great tensile strength and durability.

The word metamorphism is taken from the Greek for "change of form"; metamorphic rocks are derived from igneous or sedimentary rocks that have altered their form (recrystallized) as a result of changes in their physical environment. Metamorphism comprises changes both in mineralogy and in the fabric of the original rock. In general, these alterations are brought about either by the intrusion of hot magma into cooler surrounding rocks (contact metamorphism) or by large-scale tectonic movements of Earth's lithospheric plates that alter the pressure-temperature conditions of the rocks (regional metamorphism;). Minerals within the original rock, or protolith, respond to the changing conditions by reacting with one another to produce a new mineral assemblage that is thermodynamically stable under the new pressure-temperature conditions. These reactions occur in the solid state but may be facilitated by the presence of a fluid phase lining the grain boundaries of the minerals. In contrast to the formation of igneous rocks, metamorphic rocks do not crystallize from a silicate melt, although high-temperature metamorphism can lead to partial melting of the host rock.

Metamorphism of banded gneiss: Banded gneiss produced by metamorphism of quartzofeldspathic sediments in Scotland. White bands have undergone partial melting and recrystallized into granite. Dark bands represent material composed predominantly of biotite and minor hornblende.

Because metamorphism represents a response to changing physical conditions, those regions of Earth's surface where dynamic processes are most active will also be regions where metamorphic processes are most intense and easily observed. The vast region of the Pacific margin, for example, with its seismic and volcanic activity, is also an area in which materials are being buried and metamorphosed intensely. In general, the margins of continents and regions of mountain building are the regions where metamorphic processes proceed with intensity. But in relatively quiet places, where sediments accumulate at slow rates, less spectacular changes also occur in response to changes in pressure and temperature conditions. Metamorphic rocks are therefore distributed throughout the geologic column.

Because most of Earth's mantle is solid, metamorphic processes may also occur there. Mantlerocks are seldom observed at the surface because they are too dense to rise, but occasionally a glimpse is presented by their inclusion in volcanic materials. Such rocks may represent samples from a depth of a few hundred kilometres, where pressures of about 100 kilobars (3 million inches of mercury) may be operative. Experiments at high pressure have shown that few of the common minerals that occur at the surface will survive at depth within the mantle without changing to new, high-density phases, in which atoms are packed more closely together. Thus, the common form of SiO_2, quartz, with a density of 2.65 grams per cubic cm (1.53 ounces per cubic inch), transforms to a new phase, stishovite, with a density of 4.29 grams per cubic centimetre (2.48 ounces per cubic inch). Such changes are of critical significance in the geophysical interpretation of Earth's interior.

In general, temperatures increase with depth within Earth along curves referred to as geotherms. The specific shape of the geotherm beneath any location on Earth is a function of its corresponding local tectonic regime. Metamorphism can occur either when a rock moves from one position to another along a single geotherm or when the geotherm itself changes form. The former can take place when a rock is buried or uplifted at a rate that permits it to maintain thermal equilibrium with its surroundings. This type of metamorphism occurs beneath slowly subsiding sedimentary basins and also in the descending oceanic plate in some subduction zones. The latter process occurs when hot magma intrudes and alters the thermal state of a stationary rock or when the rock is rapidly transported by tectonic processes (e.g., thrust faulting or large-scale folding) into a new depth-temperature regime in, for example, areas of collision between two continents. Regardless of which process occurs, the result is that a collection of minerals that are thermodynamically stable at the initial conditions are placed under a new set of conditions at which they may or may not be stable. If they are no longer in equilibrium with one another under the new conditions, the minerals will react in such a way as to approach a new equilibrium state. This may involve a complete change in mineral assemblage or simply a shift in the compositions of the preexisting mineral phases. The resultant mineral assemblage will reflect the chemical composition of the original rock and the new pressure-temperature conditions to which the rock was subjected.

Because protolith compositions and the pressure-temperature conditions under which they may be placed vary widely, the diversity of metamorphic rock types is large. Many of these varieties are repeatedly associated with one another in space and time, however, reflecting a uniformity of geologic processes over hundreds of millions of years. For example, the metamorphic rock associations that developed in the Appalachian Mountains of eastern North America in response to the collision between the North American and African lithospheric plates during the Paleozoic Era (541 million to 252 million years ago) are very similar to those that developed in the Alps of south-central Europe during the collision between the European and African plates that occurred during the Mesozoic and Cenozoic eras (252 million years ago to the present). Likewise, the metamorphic rocks exposed in the Alps are grossly similar to metamorphic rocks of the same age in the Himalayas of Asia, which formed during the continental collision between the Indian and Eurasian plates. Metamorphic rocks produced during collisions between oceanic and continental plates from different localities around the world also show striking similarities to each other yet are markedly different from metamorphic rocks produced during continent-continent collisions. Thus, it is often possible to reconstruct tectonic events of the past on the basis of metamorphic rock associations currently exposed at Earth's surface.

Metamorphic Variables

Metamorphism results from a complex interplay between physical and chemical processes that operate on a scale ranging from micrometres (e.g., fine mineral grain sizes, thickness of intergranular fluid, diffusion distances for chemical species) to tens or hundreds of kilometres (e.g., crustal thickness, width of collision zone between lithospheric plates, depth to subducting plate). Despite this wide range and the many processes involved in the recrystallization of sedimentary and igneous protoliths (source rocks) into metamorphic rocks, there are relatively few variables that effect metamorphic changes. Those of greatest importance are temperature, pressure, and the original chemical composition of the protolith.

Schist and sandstone: Schist (left) is a fine-grained metamorphic rock rich in mica, and sandstone (right) is a type of sedimentary rock.

Temperature

Temperatures at which metamorphism occurs range from the conditions of diagenesis (approximately 150–200 °C) up to the onset of melting. Rocks of different compositions begin to melt at different temperatures, with initial melting occurring at roughly 650–750 °C in rocks of granitic or shaley composition and approximately 900–1,200 °C in rocks of basaltic composition. Above these temperatures, metamorphic processes gradually give way to igneous processes. Hence, the temperature realm of metamorphism spans an interval of about 150–1,100 °C and is strongly dependent on the composition of the protolith.

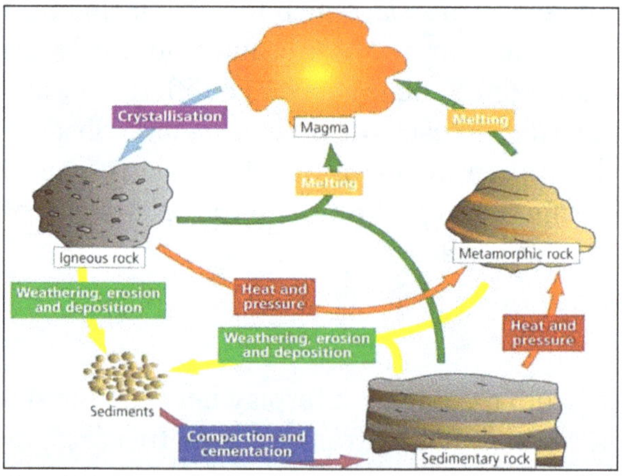

Rock cycle: The rock cycle is made up of igneous rocks, sedimentary rocks, and metamorphic rocks along with some amount of unconsolidated sediments.

The temperature at any point within Earth's crust is controlled by the local heat-flow regime, which is a composite function of heat flow upward from the mantle into the crust, heat generated by radioactive decay in nearby regions of the crust, heat transported into the crust by silicate melts, and tectonic transport of hot or cold rocks at rates faster than those needed to maintain thermal equilibrium with the surrounding rocks. The temperature gradient at any location in Earth, known as the geothermal gradient, is the increase in temperature per unit distance of depth; it is given by the tangent to the local geotherm. The magnitude of the geothermal gradient thus varies with the shape of the geotherm. In regions with high surface heat flow, such as areas of active volcanism or mantle upwelling beneath thinned continental crust, geothermal gradients of 40 to 100 °C (104 to 212 °F) per kilometre (0.6 mile) prevail, giving rise to high temperatures at relatively shallow levels of the crust. Within the stable interiors of old continents, geothermal gradients of 25 to 35 °C (77 to 95 °F) per kilometre are more typical, and in zones of active subduction, where the relatively cold oceanic crust is rapidly transported to great depths, geothermal gradients range from 10 to 20 °C (53.6 to 68 °F) per kilometre. These large variations in geotherms and geothermal gradients give rise to different metamorphic regimes, or combinations of pressure-temperature conditions, associated with the different tectonic provinces.

In addition to the variation of geotherms as a function of position in Earth, individual geotherms at a single location can vary with time. Geotherms are at steady state (i.e., do not change with time) in tectonically quiescent areas of Earth, such as the middle regions of large continents, and also in areas where tectonic processes like subduction have operated at similar rates over long periods. Transient geotherms, on the other hand, are generated in tectonically active regions, such as zones of continent-continent collision or rapid uplift and erosion, in which the tectonic processes are relatively short-lived; in these areas, the temperature at a given depth in Earth is time-dependent, and individual geotherms can have very complex shapes that with time approach smooth curves. These complex geotherms can produce wide temperature fluctuations at a given depth within Earth; rocks metamorphosed in response to these variations may show considerable textural and chemical evidence of disequilibrium, reflecting the fact that temperatures changed at rates that were more rapid than reaction rates among the constituent minerals.

Pressure

The pressure experienced by a rock during metamorphism is due primarily to the weight of the overlying rocks (i.e., lithostatic pressure) and is generally reported in units of bars or kilobars. The standard scientific notation for pressure is expressed in pascals or megapascals (1 pascal is equivalent to 10 bars). For typical densities of crustal rocks of two to three grams per cubic centimetre, one kilobar of lithostatic pressure is generated by a column of overlying rocks approximately 3.5 km (about 2 miles) thick. Typical continental crustal thicknesses are on the order of 30–40 km (roughly 19–25 miles) but can be as great as 60–80 km (about 37–50 miles) in mountain belts such as the Alps and Himalayas. Hence, metamorphism of continental crust occurs at pressures from a few hundred bars (adjacent to shallow-level intrusions) to 10–20 kilobars at the base of the crust. Oceanic crust is generally 6–10 km (about 4–6 miles) in thickness, and metamorphic pressures within the oceanic crust are therefore considerably less than in continental regions. In subduction zones, however, oceanic and, more rarely, continental crust may be carried down to depths exceeding 100 km (62 miles), and metamorphism at very high pressures may occur. Metamorphic recrystallization also occurs in the mantle at pressures up to hundreds of kilobars.

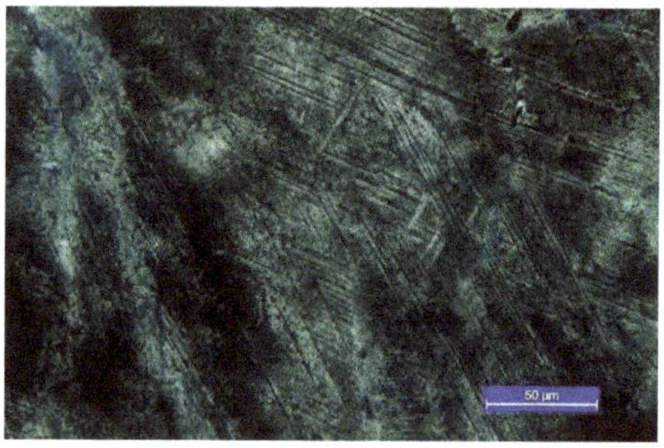

Effects of sudden pressure on rock: Unique orientations of intracrystalline features in quartz from the Lake Raeside impact crater in Leonora, Western Australia, showing the shock metamorphic pressures induced by impact.

Changes in lithostatic pressure experienced by a rock during metamorphism are brought about by burial or uplift of the sample. Burial can occur in response either to ongoing deposition of sediments above the sample or tectonic loading brought about, for example, by thrust-faulting or large-scale folding of the region. Uplift, or more properly unroofing, takes place when overlying rocks are stripped off by erosional processes or when the overburden is tectonically thinned.

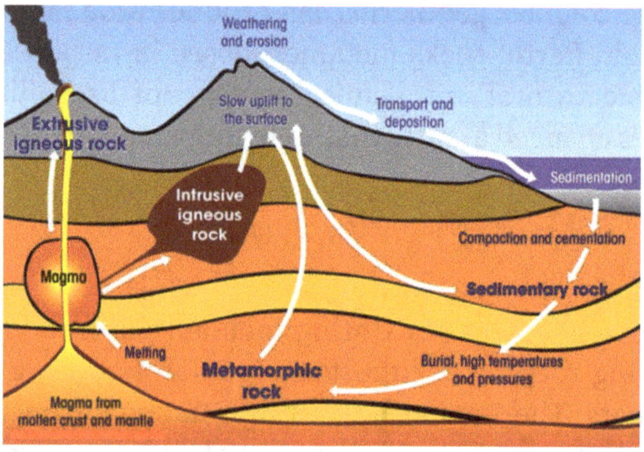

Rock cycle: The rock cycle of processes that slowly build, destroy, and alter the rocks in Earth's crust over millions of years, changing them from one form to another.

Fluids trapped in the pores of rocks during metamorphism exert pressure on the surrounding grains. At depths greater than a few kilometres within Earth, the magnitude of the fluid pressure is equal to the lithostatic pressure, reflecting the fact that mineral grain boundaries recrystallize in such a way as to minimize pore space and to seal off the fluid channelways by which solutions rise from depth. At shallow depths, however, interconnected pore spaces can exist, and hence the pressure within a pore is related to the weight of an overlying column of fluid rather than rock. Because metamorphic fluids (dominantly composed of water and carbon dioxide) are less dense than rocks, the fluid pressure at these conditions is lower than the lithostatic pressure.

Deformation of rocks during metamorphism occurs when the rock experiences an anisotropic stress—i.e., unequal pressures operating in different directions. Anisotropic stresses rarely exceed more than a few tens or hundreds of bars but have a profound influence on the textural development of metamorphic rocks.

Rock Composition

Classification into Four Chemical Systems

Common metamorphic rock types have essentially the same chemical composition as what must be their equally common igneous or sedimentary precursors. Common greenschists have essentially the same compositions as basalts; marbles are like limestones; slates are similar to mudstones or shales; and many types of gneiss are like granodiorites. In general, then, the chemical composition of a metamorphic rock will closely reflect the primary nature of the material that has been metamorphosed. If there are significant differences, they tend to affect only the most mobile (soluble) or volatile elements; water and carbon dioxide contents can change significantly, for example.

Marble: Marble is a metamorphic rock made up of granular limestone or dolomite (i.e., rock composed of calcium-magnesium carbonate) that has been recrystallized under the influence of heat, pressure, and aqueous solutions.

Despite the wide variety of igneous and sedimentary rock types that can recrystallize into metamorphic rocks, most metamorphic rocks can be described with reference to only four chemical systems: pelitic, calcareous, felsic, and mafic. Pelitic rocks are derived from mudstone(shale) protoliths and are rich in potassium (K), aluminum (Al), silicon (Si), iron (Fe), magnesium(Mg), and water (H_2O), with lesser amounts of manganese (Mn), titanium (Ti), calcium (Ca), and other constituents. Calcareous rocks are formed from a variety of chemical and detrital sediments such as limestone, dolostone, or marl and are largely composed of calcium oxide (CaO), magnesium oxide (MgO), and carbon dioxide (CO_2), with varying amounts of aluminum, silicon, iron, and water. Felsic rocks can be produced by metamorphism of both igneous and sedimentary protoliths (e.g., granite and arkose, respectively) and are rich in silicon, sodium (Na), potassium, calcium, aluminum, and lesser amounts of iron and magnesium. Mafic rocksderive from basalt protoliths and some volcanogenic sediment and contain an abundance of iron, magnesium, calcium, silicon, and aluminum. Ultramafic metamorphic rocks result from the metamorphism of mantle rocks and some oceanic crust and contain dominantly magnesium, silicon, and carbon dioxide, with smaller amounts of iron, calcium, and aluminum. For the purposes of this discussion, ultramafic rocks are considered to be a subset of the mafic category.

The particular metamorphic minerals that develop in each of these four rock chemical systems are controlled above all by the protolith chemistry. The mineral calcite ($CaCO_3$), for example, can occur only in rocks that contain sufficient quantities of calcium. The specific pressure-temperature conditions to which the rock is subjected will further influence the minerals that are produced during recrystallization; for example, at high pressures calcite will be replaced by a denser polymorph of $CaCO_3$ called aragonite. In general, increasing pressure favours denser mineral structures, whereas increasing temperature favours anhydrous and less dense mineral phases. Although some minerals, such as quartz, calcite, plagioclase, and biotite, develop under a variety of conditions, other minerals are more restricted in occurrence; examples are lawsonite, which is produced primarily during high-pressure, low-temperature metamorphism of basaltic protoliths, and sillimanite, which develops during relatively high-temperature metamorphism of pelitic rocks.

Thermodynamics of Metamorphic Assemblages

The number of minerals present in an individual metamorphic rock is limited by the laws of thermodynamics. The number of mineral phases that can coexist stably in a metamorphic rock at a particular set of pressure-temperature conditions is given by the Gibbs phase rule: number of mineral phases = number of chemical components – number of degrees of freedom + 2, where the 2 stands for the two variables of pressure and temperature. The degrees of freedom of the system are the parameters that can be independently varied without changing the mineral assemblage of the rock. For example, a rock with no degrees of freedom can only exist at a single set of pressure-temperature conditions; if either the pressure or the temperature is varied, the minerals will react with one another to change the assemblage. A rock with two degrees of freedom can undergo small changes in pressure or temperature or both without altering the assemblage. Most metamorphic rocks have mineral assemblages that reflect two or more degrees of freedom at the time the rock recrystallized. Thus, a typical pelitic rock made up of the six chemical components silica (SiO_2), aluminum oxide (Al_2O_3), ferrous oxide (FeO), magnesium oxide (MgO), potash (K_2O), and water would contain no more than six minerals; the identity of those minerals would be controlled by the pressure and temperature at which recrystallization occurred. In such a rock taken from Earth's surface, the identity of the six minerals could be used to infer the approximate depth and temperature conditions that prevailed at the time of its recrystallization. Rocks that contain more mineral phases than would be predicted by the phase rule often preserve evidence of chemical disequilibrium in the form of reactions that did not go to completion. Careful examination of such samples under the microscope can often reveal the nature of these reactions and provide useful information on how pressure and temperature conditions changed during the burial and uplift history of the rock.

Metamorphic rocks only rarely exhibit a chemical composition that is characteristically "metamorphic." This statement is equivalent to saying that diffusion of materials in metamorphism is a slow process, and various chemical units do not mix on any large scale. But occasionally, particularly during contact metamorphism, diffusion may occur across a boundary of chemical dissimilarity, leading to rocks of unique composition. This process is referred to as metasomatism. If granite is emplaced into a limestone, the contact region may be flooded with silica and other components, leading to the formation of a metasomatic rock. Often such contacts are chemically zoned. A simple example is provided by the metamorphism of magnesium-rich igneous rocks in contact with quartz-rich sediments. A zonation of the type serpentine-talc-quartz may be found, such as $Mg_6(Si_4O_{10})(OH)_8$–$Mg_3(Si_4O_{10})(OH)_2$–SiO_2.

In this case the talc zone has grown by silica diffusion into the more silica-poor environment of the serpentine. Economic deposits are not uncommon in such situations—e.g., the formation of the $CaWO_4$ (calcium tungstate) scheelite when tungstate in the form of WO_3 moves from a granite into a limestone contact. The reaction can be expressed as $CaCO_3 + WO_3$ (solution) $\rightarrow CaWO_4 + CO_2$ (gas).

Metamorphic Reactions

Reactions in a Kaolinite-quartz System

A very simple mineralogical system and its response to changing pressure and temperature provide a good illustration of what occurs in metamorphism. An uncomplicated sediment at Earth's surface, a mixture of the clay mineral kaolinite [$Al_4Si_4O_{10}(OH)_8$] and the mineral quartz(SiO_2), provides a good example. Most sediment has small crystals or grain sizes but great porosity and permeability, and the pores are filled with water. As time passes, more sediment are piled on top of the surface layer, and it becomes slowly buried. Accordingly, the pressure to which the layer is subjected increases because of the load on top, known as overburden. At the same time, the temperature will increase because of radioactive heating within the sediment and heat flow from deeper levels within Earth.

In the first stages of incremental burial and heating, few chemical reactions will occur in the sediment layer, but the porosity decreases, and the low-density pore water is squeezed out. This process will be nearly complete by the time the layer is buried by 5 km (3 miles) of overburden. There will be some increase in the size of crystals; small crystals with a large surface area are more soluble and less stable than large crystals, and throughout metamorphic processes there is a tendency for crystals to grow in size with time, particularly if the temperature is rising.

Eventually, when the rock is buried to a depth at which temperatures of about 300 °C (572 °F) obtain, a chemical reaction sets in, and the kaolinite and quartz are transformed to pyrophylliteand water: $Al_4Si_4O_{10}(OH)_8 + 4SiO_2 \rightarrow 2Al_2Si_4O_{10}(OH)_2 + 2H_2O$.

The exact temperature at which this occurs depends on the fluid pressure in the system, but in general the fluid and rock-load pressures tend to be rather similar during such reactions. The water virtually fights its way out by lifting the rocks. Thus, the first chemical reaction is a dehydration reaction leading to the formation of a new hydrate. The water released is itself a solvent for silicates and promotes the crystallization of the product phases.

If heating and burial continue, another dehydration sets in at about 400 °C (752 °F), in which the pyrophyllite is transformed to andalusite, quartz, and water: $Al_2Si_4O_{10}(OH)_2 \rightarrow Al_2SiO_5 + 3SiO_5 + H_2O$.

After the water has escaped, the rock becomes virtually anhydrous, containing only traces of fluid in small inclusions in the product crystals and along grain boundaries. Both of these dehydration reactions tend to be fast, because water, a good silicate solvent, is present.

Although the mineral andalusite is indicated as the first product of dehydration of pyrophyllite, there are three minerals with the chemical composition Al_2SiO_5. Each has unique crystal structures, and each is stable under definite pressure-temperature conditions. Such differing forms with identical composition are called polymorphs. If pyrophyllite is dehydrated under high-pressure conditions, the polymorph of Al_2SiO_5 formed would be the mineral kyanite (the most dense polymorph). With continued heating, the original andalusite or kyanite will invert to sillimanite, the highest-temperature Al_2SiO_5 polymorph:

The kinetics of these polymorphic transformations are sufficiently sluggish, however, that kyanite or andalusite may persist well into the stability field of sillimanite.

Reactions of other Mineral Systems

Because of the very simple bulk composition of the protolith in this example (a subset of the pelitic system containing only SiO_2-Al_2O_3-H_2O), no other mineralogical changes will occur with continued heating or burial. The original sediment composed of kaolinite, quartz, and water will thus have been metamorphosed into a rock composed of sillimanite and quartz and perhaps some metastable andalusite or kyanite, depending on the details of the burial and heating history. In the case of a more typical pelite containing the additional chemical components potash, ferrous oxide, and magnesium oxide, the reaction history would be correspondingly more complex. A typical shale that undergoes burial and heating in response to continent-continent collision would develop the minerals muscovite, chlorite, biotite, garnet, staurolite, kyanite, sillimanite, and alkali feldspar, in approximately that order, before beginning to melt at about 700 °C (1,292 °F). Each of these minerals appears in response to a chemical reaction similar to those presented above. Most of these reactions are dehydration reactions, and the shale thus loses water progressively throughout the entire metamorphic event. As discussed above, the total number of minerals present in the rock is controlled by the Gibbs phase rule, and the addition of new minerals generally results from the breakdown of old minerals. For example, the reaction involving garnet, muscovite, and chlorite that yields staurolite, biotite, quartz, and water occurs at temperatures of about 500–550 °C (932–1,022 °F) and typically consumes all the preexisting chlorite in the rock, introduces the new mineral staurolite, and adds more biotite and quartz to the biotite and quartz generated by earlier reactions. Some garnet and muscovite usually remain after the reaction, although examination of the sample under the microscope would probably reveal partial corrosion (wearing away due to chemical reactions) of the garnets resulting from their consumption.

Isograds

Reactions that introduce new minerals in rocks of a specific bulk composition are referred to as mineral appearance isograds. Isograds can be mapped in the field as lines across which the metamorphic mineral assemblage changes. Caution must be exercised to note the approximate bulk composition of the rocks throughout the map area, however, because the same mineral can develop at quite different sets of pressure-temperature conditions in rocks of dissimilar chemical composition. For example, garnet generally appears at a lower temperature in pelitic rocks than it does in mafic rocks; hence, the garnet isograd in pelitic rocks will not be the same map line as that in metabasaltic (i.e., metamorphosed basalt) compositions.

Principal Types

Metamorphic reactions can be classified into two types that show different degrees of sensitivity to temperature and pressure changes: net-transfer reactions and exchange reactions. Net-transfer reactions involve the breakdown of preexisting mineral phases and corresponding nucleation and growth of new phases. (Nucleation is the process in which a crystal begins to grow from one or more points, or nuclei.) They can be either solid-solid reactions (mineral A + mineral B = mineral C + mineral D) or devolatilization reactions (hydrous mineral A = anhydrous mineral B + water), but in either case they require significant breaking of bonds and reorganization of material in the rock. They may depend most strongly on either temperature or pressure changes. In general, devolatilization reactions are temperature-sensitive, reflecting the large increase in entropy (disorder) that accompanies release of structurally bound hydroxyl groups (OH^-) from minerals to produce molecular water. Net-transfer reactions that involve a significant change in density of the participating mineral phases are typically more sensitive to changes in pressure than in temperature. An example is the transformation of albite ($NaAlSi_3O_8$) to the sodic pyroxene jadeite ($NaAlSi_2O_6$) plus quartz (SiO_2). Albite and quartz have similar densities, of about 2.6 grams per cubic cm (1.5 ounces per cubic inch), whereas jadeite has a density of 3.3 grams per cubic cm (1.9 ounces per cubic inch). The increased density reflects the closer packing of atoms in the jadeite structure. Not surprisingly, the denser phase jadeite is produced during subduction zone (high-pressure) metamorphism. Net-transfer reactions always involve a change in mineral assemblage, and textural evidence of the reaction often remains in the sample; isograd reactions are invariably net-transfer reactions.

In contrast to net-transfer reactions, exchange reactions involve redistribution of atoms between preexisting phases rather than nucleation and growth of new phases. The reactions result simply in compositional changes of minerals already present in the rock and do not modify the mineral assemblage. For example, the reaction involving almandine garnet($Fe_3Al_2Si_3O_{12}$) and magnesium biotite ($KMg_3AlSi_3O_{10}(OH)_2$) that yields pyrope garnet ($Mg_3Al_2Si_3O_{12}$) and iron biotite ($KFe_3AlSi_3O_{10}(OH)_2$) results in redistribution of iron and magnesium between garnet and biotite but creates no new phases. This reaction is limited by the rates at which iron and magnesium can diffuse through the garnet and biotite structures. Because diffusion processes are strongly controlled by temperature but are nearly unaffected by pressure, exchange reactions are typically sensitive to changes only in metamorphic temperature. Exchange reactions leave no textural record in the sample and can be determined only by detailed microanalysis of the constituent mineral phases. The compositions of minerals as controlled by exchange reactions can provide a useful record of the temperature history of a metamorphic sample.

The types of reactions cited here are typical of all metamorphic changes. Gases are lost (hydrous minerals lose water, carbonates lose carbon dioxide), and mineral phases undergo polymorphic or other structural changes; low-volume, dense mineral species are formed by high pressures, and less dense phases are favoured by high temperatures. Considering the immense chemical and mineralogical complexity of Earth's crust, it is clear that the number of possible reactions is vast. In any given complex column of crustal materials some chemical reaction is likely for almost any incremental change in pressure and temperature. This is a fact of immense importance in unraveling the history and mechanics of Earth, for such changes constitute a vital record and are the primary reason for the study of metamorphic rocks.

Retrograde Metamorphism

In general, the changes in mineral assemblage and mineral composition that occur during burial and heating are referred to as prograde metamorphism, whereas those that occur during uplift and cooling of a rock represent retrograde metamorphism. If thermodynamic equilibrium were always maintained, one might expect all the reactions that occur during prograde metamorphism to be reversed during subsequent uplift of the rocks and reexposure at Earth's surface; in this case, metamorphic rocks would never be seen in outcrop. However, two factors mitigate against complete retrogression of metamorphic rocks during their return to Earth's surface. First is the efficient removal of the water and carbon dioxide released during prograde devolatilization reactions by upward migration of the fluid along grain boundaries and through fractures. Because almost all the water released during heating by reactions—such as when chlorite ($Fe_9Al_6Si_5O_{20}(OH)_{16}$) reacts with quartz ($4SiO_2$) to yield garnet ($3Fe_3Al_2Si_3O_{12}$) and water ($8H_2O$)—is removed from the site of reaction, the reaction cannot be reversed during cooling unless water is subsequently added to the rock. Thus, garnet can be preserved at Earth's surface even though it is thermodynamically unstable at such low temperatures and pressures. The second reason that metamorphic reactions do not typically operate in reverse during cooling is that reaction rates are increased by rising temperatures. During cooling, reaction kinetics become sluggish, and metastable mineral assemblages and compositions can be preserved well outside their normal stability fields. Thus, prograde reactions are generally more efficient than retrograde reactions and metamorphic assemblages indicative of even extremely high temperatures or pressures or both are found exposed throughout the world.

It is common, however, to find at least some signs of retrogression in most metamorphic rocks. For example, garnets are often rimmed by small amounts of chlorite and quartz, indicating that limited quantities of water were available for the reverse of the reaction given above to proceed during cooling. Retrograde features such as these reaction rims can be mapped to yield information on pathways of fluid migration through the rocks during uplift and cooling. In other rocks, such as high-temperature gneisses, mineral compositions often reflect temperatures too low to be in equilibrium with the preserved mineral assemblage. In these samples, it is clear that certain exchange reactions operated in a retrograde sense even when the net-transfer reactions were frozen in during prograde metamorphism.

Textural Features

The fabric of a metamorphic rock results from the combined effects of mineral reactions and deformation throughout the metamorphic event and the subsequent return of the rock to the terrestrial surface. The study of metamorphic fabrics in outcrop and under the microscope has become a highly specialized subject aimed at revealing the nature and direction of the forces acting during dynamic processes within Earth. Much of this work is an outgrowth of a classic investigation conducted in 1930 by the Austrian geologist Bruno Sander, coupled with more recent experimental work on the pressure-temperature stabilities of metamorphic minerals and their responses to deformation.

Observations show that pressure is only rarely hydrostatic (equal in all directions) at any point within Earth's crust. In real cases, consequently, anisotropic stresses operate that may lead to flow or fracture of materials. Such occurrences produce certain characteristic fabrics or structures in metamorphic rocks that may be observed at the scale of the orientation of small crystals in a rock or as a pattern of folds in a mountain range. One of the principal characteristics of most metamorphic rocks is that the arrangement of crystals is not isotropic, or random, but that there is a strong preferred orientation related to the direction of stress components of pressure. Such preferred orientation of crystals and mineral grains is perhaps the most striking difference between metamorphic rocks and other rock types.

Major Features

The most obvious features of metamorphic rocks are certain planar features that are often termed s-surfaces. The simplest planar features may be primary bedding (akin to the layering in sedimentary rocks). As the rock crystallizes or recrystallizes under directed pressure, new crystals may grow in some preferred direction, sometimes subparallel to the primary bedding but often at new angles defining new planar structures. At the same time, folding of layers may occur, leading to folds with amplitudes on scales of kilometres or millimetres. Fabric symmetry may be represented by the nature of deformed fossils, pebbles in a conglomerate, or any objects with a known shape prior to deformation.

A few terms that commonly are used to describe several types of preferred orientation in metamorphic rocks include foliation, a general term describing any type of s-surface, bedding, or crystal orientation; slaty cleavage, a planar structure leading to facile cleavage that is normally caused by the preferred orientation of mica crystals; schistosity, a term used to describe repetitive and pronounced foliation of the type that is present in schists; and lineation, which is any linear structure, such as the axis of the fold, grooves on a fault plane, or the direction of stretching of pebbles.

The various mineral phases of a metamorphic rock have different physical properties and symmetries. When a rock is subjected to recrystallization in a stress field, different substances will behave differently according to such physical properties and symmetries. Some minerals always tend to grow in better-formed crystals than others; rates of nucleation may differ, and this can lead to different patterns of growth of crystals—there may be a few large crystals or a mass of small crystals. Minerals can be arranged in order of their tendency to form crystals showing planar surfaces—namely, magnetite, garnet, epidote, mica, calcite, quartz, and feldspar. Minerals that have a tendency to form large single crystals (e.g., garnet) are termed porphyroblasts.

Porphyroblastic crystals may grow before, during, or after an episode of deformation (pre-, syn-, or postkinematic growth, respectively); the relative timing of mineral growth and deformation can often be determined by examining the sample under a microscope. Prekinematic porphyroblasts may be fractured by subsequent deformation; the orientation of the fractures and any offset of the grains along them provide information on the directed stresses responsible for the deformation. Prekinematic grains may also be surrounded by pressure shadows produced by minerals such as quartz that dissolve in zones of maximum compressive stress and reprecipitate in zones of lesser compressive stress adjacent to the rigid porphyroblasts. The texture of the shadows is different from that of the host rock. Samples exhibiting asymmetric pressure shadows around porphyroblasts can yield information on the orientation of shear stresses during deformation. A spectacular example of synkinematic prophyroblast growth—that is, recrystallization resulting in the development of a secondary crystal—is provided by the so-called snowball garnets, which have spiral trails of inclusions that indicate rotation of either the garnet or the matrix foliation during garnet growth. Postkinematic porphyroblasts typically overgrow all previous fabrics in the rock and may contain trails of mineral inclusions that define microfolds or an earlier schistosity.

In some samples, it is possible to use the compositions of the porphyroblasts to calculate the depth and temperature conditions at which they grew and thereby constrain the conditions at which deformation occurred. Studies of this sort add immeasurably to the understanding of crustal rheologies and the response of rocks to large-scale orogenic events. Because a particular metamorphic event may be accompanied by either several isolated episodes of deformation or a single continuum of deformation, there may be many fabric generations recorded in one sample; individual minerals may be postkinematic with respect to the earliest deformation but prekinematic relative to younger deformation in the same rock. Thus, the study of porphyroblast fabrics in metamorphic rocks can be complex but has the potential to yield important information on the structural history of metamorphic regions.

Because changes in pressure and temperature often occur at faster rates than those of mineral reaction and recrystallization, metamorphic rocks may display fabrics that result from incomplete reactions. Such disequilibrium features provide a wealth of information on the reaction history of the sample and, by comparison with experimental studies of mineral stabilities, can also constrain the quantitative pressure-temperature history of the rock during metamorphism.

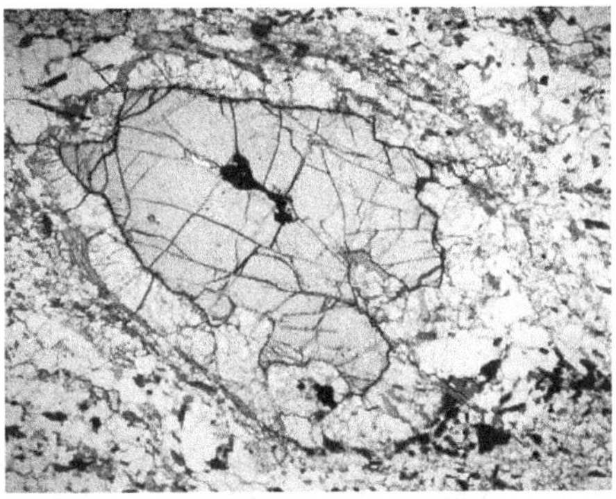

Photomicrograph showing corroded garnet (gray) surrounded by a corona of cordierite produced during uplift of the sample. Other minerals present are biotite, plagioclase, sillimanite, alkali feldspar, and ilmenite. The garnet is two millimetres across.

Reaction textures occur at the corroded margins of crystals, from the corrosive rimming of crystals of one mineral by finer-grained aggregates of another or as a result of other features that indicate partial removal of crystalline material by reaction with magma or an acid. For example, a corroded garnet may be surrounded by a corona (reaction rim) of the mineral cordierite. Other minerals present in the matrix could include sillimanite, quartz, biotite, and alkali feldspar. These textural features suggest the following reaction relationship between garnet, sillimanite, quartz, and cordierite: garnet ($2Mg_3Al_2Si_3O_{12}$) + sillimanite ($4Al_2SiO_5$) ⇌ cordierite ($3Mg_2Al_4Si_5O_{18}$).

Such a reaction has been shown experimentally to occur at temperatures of approximately 725 ± 50 °C (1,337 ± 122 °F) and to be very sensitive to pressure, with cordierite occurring under low-pressure conditions. The textural evidence that preexisting garnet was partially replaced by cordierite thus implies that the rock underwent decompression while still at high temperatures and that the decompression occurred too rapidly for the rock to recrystallize completely (i.e., for garnet to be totally replaced by cordierite).

Lamination

There is also a tendency for many types of metamorphic rocks to become laminated (that is, to accumulate in successive layers, or laminae), and the separate laminae may have distinct chemical compositions. A macroscopically rather homogeneous sediment may prove to be inhomogeneous on a minute scale. When graywackes are metamorphosed within the greenschist facies, for example, laminae rich in quartz and feldspar alternate with others rich in epidote, chlorite, and muscovite. The precise causes of this process are not well known, but it may result from a combination of extensive deformation accompanied by recrystallization. In a sense, it is a type of flow unmixing. It is important to recognize that this type of structure need have no relation to original bedding in the unmetamorphosed sediments.

Structural Features

Metamorphic rocks are often intimately related to large-scale (kilometres of tens of kilometres) structural features of Earth. Such features include folds, nappes, and faults with a wide variety of geometries. In many cases, the correlation of metamorphic isograds and their position in the structure implies a genetic relationship between the two. For example, one of the major structural features in the Himalayan mountain belt is the Main Central Thrust, a thrust fault that runs for hundreds of kilometres from east to west and was responsible for the transportation of rocks belonging to the Eurasian Plate southward over those of the Indian Plate. Along much of the length of this fault, the metamorphic rocks in the hanging wall (located above the fault) display a pattern of inverted isograds; i.e., the rocks that reached the highest temperatures of metamorphism overlie rocks that record lower temperatures, implying that metamorphic temperatures decreased with depth to the fault. Several explanations have been proposed to account for this anomalous distribution of temperature with depth. One model suggests the fault transported hot Asian rocks over cooler Indian rocks, which caused cooling of the Asian rocks in the vicinity of the fault. Another model proposes that fluids circulating along the fault zone caused retrograde metamorphism and thus reset the rocks located nearest to the fault to lower temperatures. Although neither of these models provides an adequate explanation for the entire length of the Main Central Thrust, they both emphasize the significant control that structural features can exert on the development of metamorphic rocks.

Himalayas: The Himalayas, northern Nepal.

Metamorphism associated with nappes (large recumbent folds) in the Alps and the Appalachians provides strong evidence that the tectonic transport of rocks typically occurs at rates faster than those of thermal equilibration—in other words, that the nappes can transport hot rocks for large distances without significant cooling. Nappe formation is a major process of crustal thickening during continent-continent collision; emplacement of the nappes results in burial and heating of the underlying rocks. Isograd distributions associated with nappe structures can be either normal or inverted, depending on the relative rates of nappe emplacement and heat transfer.

Isograd maps can provide information on the relative timing of structural and metamorphic events in much the same way that fabric studies constrain the relative timing of deformation episodes and prophyroblast growth. For example, isograd patterns that are cut by faults clearly indicate that metamorphism predated fault displacement, whereas isograd sequences that overprint structural discontinuities imply the reverse. Isograds that parallel major structures suggest some cause-and-effect relationship between the structural and metamorphic development of the region. Since the 1980s, metamorphic petrologists and structural geologists have increasingly worked together to correlate metamorphic and tectonic events and thereby increase understanding of crustal dynamics in tectonically active regions of Earth.

Metamorphic Facies

Metamorphic petrologists studying contact metamorphism early in the 20th century introduced the idea of metamorphic facies (part of a rock or group of rocks that differs from the whole formation) to correlate metamorphic events. The concept was first defined in 1914 by a Finnish petrologist, Pentti Eelis Eskola, as any rock of a metamorphic formation that has attained chemical equilibrium through metamorphism at constant temperature and pressure conditions, with its mineral composition controlled only by the chemical composition. In current usage, a metamorphic facies is a set of metamorphic mineral assemblages, repeatedly associated in space and time, such that there is a constant and therefore predictable relation between mineral composition and chemical composition.

The facies concept is more or less observation-based. In a single outcrop, for instance, layers of different chemical composition will display different mineral assemblages despite having all experienced the same pressure and temperature history. A pelitic layer (that is, a layer made up of mud or clay particles) might contain the assemblage garnet + chlorite + biotite + muscovite + quartz, whereas a basaltic horizon a few centimetres away would contain the assemblage chlorite + actinolite + albite. Both of these rocks belong to the same facies, meaning that, in another region, a geologist who observed the assemblage chlorite + actinolite + albite in a metabasalt could predict that associated pelitic rocks would contain the garnet + chlorite + biotite + muscovite + quartz assemblage.

Experimental work on the relative stabilities of metamorphic minerals and assemblages has permitted correlation of the empirically derived facies with quantitative pressure and temperature conditions. The names of metamorphic facies in common usage are derived from the behaviour of a rock of basaltic bulk composition during metamorphism at various sets of pressure-temperature conditions. For example, basalt metamorphosed during subduction to high pressures at low temperatures recrystallizes into a rock containing glaucophane, lawsonite, and albite; glaucophane is a sodic amphibole that is blue to black in hand sample and lavender to blue under the microscope. Because of their distinctive bluish coloration, such samples are called blueschists. The same rock type metamorphosed at more moderate pressures and temperatures in the range of 400–500 °C (752–932 °F) would contain abundant chlorite and actinolite, minerals that are green both in hand sample and under the microscope, and would be referred to as a greenschist. At somewhat higher temperatures, the rock would become an amphibolite, reflecting mineralogy composed predominantly of the amphibole hornblende along with plagioclase and perhaps some garnet. At still higher temperatures, a metabasalt recrystallizes into a rock containing hypersthene, diopside, and plagioclase; in general, these minerals form relatively equant crystals and hence do not develop a preferred orientation. The granular texture of these rocks has resulted in the name granulite for a high-temperature metabasalt. A pelitic or calcareous rock will develop very different mineral assemblages from a metabasalt, yet the same facies names apply. Thus, one can refer to a greenschist facies pelitic schist, an amphibolite facies calcsilicate rock, or a granulite facies garnet gneiss.

The boundaries between the different facies are regions of pressure and temperature in which chemical reactions occur that would significantly alter the mineralogy of a rock of basaltic bulk composition. For example, the boundary between the greenschist and amphibolite facies marks a transition from amphibole of actinolitic composition to hornblende and of a sodic plagioclase into a more calcic plagioclase. The reactions that bring about these transformations depend on the specific composition of the rock.

Facies Series

Different types of tectonic processes produce different associations of metamorphic facies in the field. For example, regions associated with subduction of oceanic material beneath either oceanic crust or continental crust are characterized by blueschist, greenschist, and eclogite facies rocks, whereas areas thought to reflect continent-continent collision are more typically distinguished by greenschist and amphibolite facies rocks. Still other regions, usually containing an abundance of intrusive igneous material, show associations of low-pressure greenschist, amphibolite, and granulite facies rocks. These observations led a Japanese petrologist, Akiho Miyashiro, working in the 1960s and '70s, to develop the concept of baric types, or metamorphic facies series. Miyashiro described the three facies associations given above as high-pressure, medium-pressure, and low-pressure facies series, respectively, and correlated the development of these characteristic series with the shape of the geotherm (a line or surface connecting points of equal temperature either on or within Earth) in different tectonic settings. Subsequent thermal modeling studies have shown that metamorphism generally occurs in response to tectonically induced perturbation of geotherms rather than along steady-state geotherms and, hence, that the facies series do not record metamorphic geotherms. Nonetheless, the concept of metamorphic facies series is a useful one in that it emphasizes the strong genetic relationship between metamorphic style and tectonic setting.

Pressure-temperature-time Paths

Interaction between metamorphic petrologists and geophysicists in the 1980s led to the realization that each metamorphic rock follows its own unique path through pressure-(depth-) temperature space during metamorphism and that these paths bear little or no resemblance to steady-state geotherms. The specific shape of a pressure-temperature-time (P-T-t) path depends on the tectonic setting in which the rock was metamorphosed, which in turn controls the relative rates at which burial or uplift and heating or cooling occur. For example, a rock that is uplifted rapidly from depth, perhaps in response to extensional faulting (that caused by the stretching of Earth's crust), may transport heat with it to near-surface depths. Its P-T-t path would show a phase of nearly isothermal decompression (uplift at approximately constant temperature), reflecting the fact that uplift rates were more rapid than those of heat transfer. In contrast, if the same rock remained at depth for a long period of time and then experienced very slow uplift, its P-T-t path would show cooling during uplift or even a phase of isobaric (constant-pressure) cooling. Rocks belonging to medium-pressure facies series generally follow P-T-t paths that are clockwise loops on a pressure-temperature diagram, reflecting rapid burial during a collisional event followed by heating and relatively rapid uplift.

In contrast, low-pressure facies series rocks may follow counterclockwise P-T-t paths in response to rapid heating of the crust due to magma intrusion prior to uplift. P-T-t paths followed by rocks of a high-pressure facies series are less predictable and depend strongly on the mechanism by which the rocks are transferred from the subducting slab to the overlying continental crust. In general, the mineral assemblage preserved in a metamorphic rock is frozen at the highest temperature experienced during metamorphism, and thus the facies and facies series to which the rock would be assigned reflect only a single point on its P-T-t path.

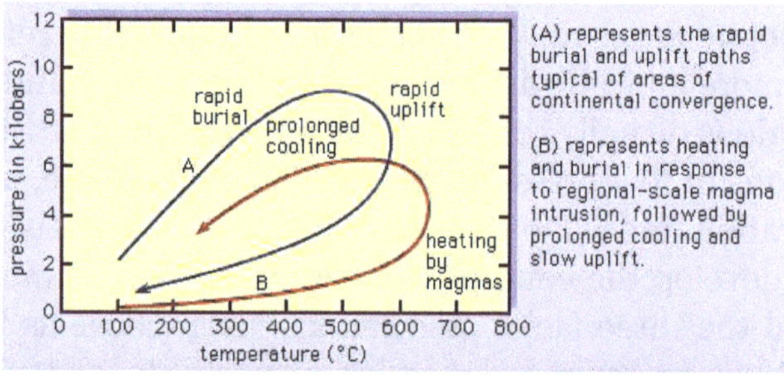

Metamorphic rock: Pressure-temperature-time paths: Examples of pressure-temperature-time paths followed by rocks metamorphosed in different environments.

One of the principal goals of much of the work that is done on metamorphic rocks is the reconstruction of the P-T-t paths followed by rocks presently exposed at Earth's surface. Because these paths are so strongly linked to dynamic processes, their reconstruction provides a means by which tectonic processes operative in the geologic past may be understood. Owing to the continuous recrystallization of rocks that occurs during progressive metamorphism, much of the early record of metamorphic changes within a sample is eradicated by later events. It is, therefore, not possible to determine the entire P-T-t path followed by an individual sample, but often enough disequilibrium features are preserved to permit reconstruction of a few thousand bars and a couple of hundred degrees of the path; such a portion may represent anywhere from a few million to a hundred million years of Earth history, as revealed by geochronologic determinations involving different minerals or fabric generations in the sample. Techniques for determining the pressure-temperature history of a metamorphic rock include using compositions of coexisting minerals to calculate pressures and temperatures of equilibration (geobarometry and geothermometry, respectively), comparing the mineral assemblage to experimentally determined stability fields for the phases, utilizing mineral inclusions enclosed within porphyroblasts to constrain assemblages present in the early history of the sample, and making use of the densities of small inclusions of fluids trapped within the minerals to determine possible pressures and temperatures experienced at different stages in the burial and uplift history.

Distribution of Metamorphic Rocks

A high-grade metamorphic rock is one that formed at a depth of tens of kilometres and later returned to the surface. Hence, metamorphic regions are also regions of former or recent intense orogeny. More-stable regions of Earth's crust tend to be covered with sediments, and only deep drilling will reveal the metamorphic rocks below.

Earth's crust is made up of two basic units, the continents and ocean basins. Exploration of ocean floors has revealed that old, thick sedimentary piles are missing. Doubtless this is related to the processes of continental drift or seafloor spreading; sediments are continuously swept up by continental motion and are added to the continents or returned to the upper mantle. Nearly all studies of metamorphic rocks have concentrated on the continents for this reason.

There are few large areas of Earth's crust that are not affected by some type of igneous event from time to time. Although the intensity of volcanism may be focused in certain geographic regions (e.g., the Pacific margin), volcanism appears to be a rather random phenomenon, at times even occurring in the stable shield areas of the continents. In this sense, contact-metamorphic events may be found almost everywhere at almost any time on Earth. But these metamorphic events are of trivial volumetric significance compared with those of regional metamorphism.

During the past 500 million years or so of Earth history, major tectonic, seismic, igneous, and metamorphic events have been concentrated on continental margins. This has been a period of depression and uplift of Earth's crust associated with the formation of the present continental distribution. The processes are still going on at dramatic rates in ocean trench environments. These modern regions of activity form immense linear belts. One such belt runs around virtually the entire Pacific margin and another through the Mediterranean and southern Asia to fuse with the circum-Pacific belt. It is in these belts that the spectacular development of zeolite facies, prehnite-pumpellyite facies, blueschist facies, and, occasionally, eclogite facies, as well as the more universal facies of regional metamorphism, have occurred. The granulite facies is almost missing.

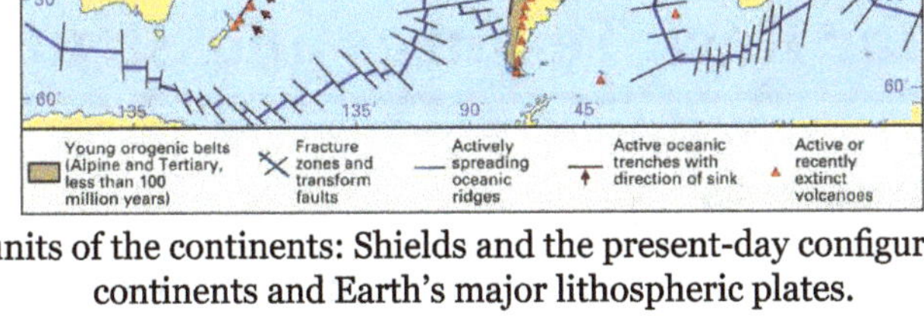

Tectonic units of the continents: Shields and the present-day configuration of the continents and Earth's major lithospheric plates.

The central and often dominant feature of most continents is their vast Precambrian shield area; examples include the Canadian Shield, Brazilian Shield, African Shield, and Australian Shield. In these rocks, dating reveals ages of 1 billion to 4.28 billion years, and they have been little affected by tectonic events postdating the Cambrian. But these shield areas are themselves complex. They consist of vast areas of granitic or granodioritic gneisses. Inside them, between them, and overlapping onto them are belts of sedimentary rocks quite like those in modern sedimentary belts of the Pacific margin or the European Alps. These rocks are frequently metamorphosed in the greenschist, amphibolite, and granulite facies. Low-temperature facies and, in particular, low-temperature–high-pressure facies are missing—or have not yet been found. From marginal areas of these stable shield areas a complex array of processes has been documented covering the past few hundred million years. The Caledonian orogeny (at the close of the Silurian Period) produced tectonic-metamorphic events along the east coast of North America, Greenland, the British Isles, Fennoscandia (the region made up of Scandinavia, Finland, and northwestern Russia), Central Asia, and Australia. The Hercynian, or Variscan, orogeny followed about 300 million years ago, affecting subparallel regions and the Urals and European Alps. In fact, the shield margins appear to have been subjected to a more or less constant battering by forces both destroying and rebuilding the margins of these protocontinents. As geologists study Precambrian areas in greater detail, the number of metamorphic and orogenic events recognized on a global scale increases.

It is the great task and problem of those who study metamorphic rocks to deduce from them the record of Earth dynamics and thermal history. Among the questions to be answered are (1) whether the pattern of facies development through time—e.g., the granulite facies in the Archean Eon (4 billion to 2.5 billion years ago) to blueschist facies in the early Cenozoic—is a reflection of a cooling Earth and the decline of radioactivity in the crust and (2) whether the increase in size of global tectonic-metamorphic belts through time reflects changes in convective patterns in the mantle.

As understanding of the pressure-temperature regimes of metamorphism increases, and as knowledge of rock mechanics and fluid motion during metamorphism also increases through field and laboratory studies, it may become possible to understand the details of the motion of the chemical elements during such processes and hence much of the subject of economic geology, or the search for essential raw materials.

Metamorphic Rock Textures and Structures

Metamorphic Structures

Definite mechanical conditions and recrystallization determine the metamorphic structures. Five major types of metamorphic structures have been recognised as follows:

1. Cataclastic structure: This type of structure is produced under stress and in absence of high temperature, whereby rocks are subjected to shearing and fragmentation. Only

the durable mineral partly survive the crushing force and the less durable ones are powdered. Thus, when resistant minerals and rock fragments stand out in a pseudo porphyritic manner in the finer materials, it is known as 'porphyroclastic structure.' Phenocrysts are called 'porphyroclasts. Argillaceous rocks develop salty cleavage, harder rocks may be shattered and crushed forming crush breccia and crush conglomerate. When the rocks are highly crushed into fine grained rocks, they are known as mylonites. Since these structures are formed due to cataclasis, they are, as a whole, known as cataclastic structure.

2. Maculose structure: This type of structure is produced by thermal metamorphism of argillaceous rocks like shales. Here, larger crystals of andalusite, cordierite and biotite are sometimes well developed giving a spotted appearance to the rocks. The well-developed crystals are known as 'porphyroblasts' with increasing degree of metamorphism, the spotted slates pass into extremely fine grained granular rock known as Hornfels.

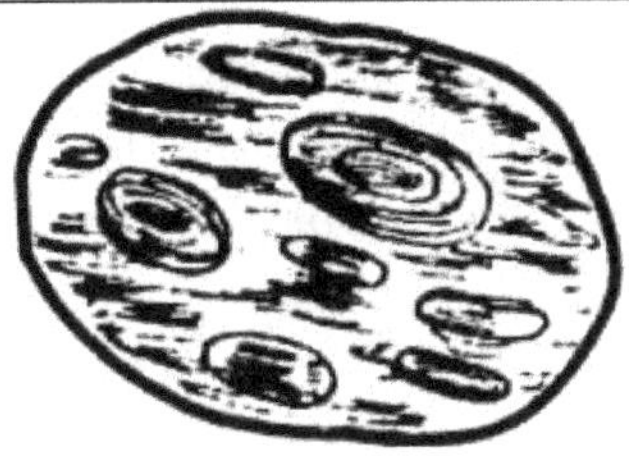

Maculose Structure(Hornfels)

3. Schistose structure: This type of structure is when the platy or flaky minerals like the micas and other inequidimensional minerals show a preferred orientation along parallel planes, under the effect of the stress dominating during metamorphism. The longer directions are parallel to the direction of maximum stress. Schistosity is the property or tendency of a foliated rock, whereby it can be readily split along foliation plane.

Schistose Structure (Mica Schist)

4. Granulose structure: This is found in the rocks composed of equidimensional minerals like quartz, feldspar and pyroxenes. They are formed by the recrystallization of pre-existing rocks, under uniform pressure and great heat. The typical texture is

coarsely granoblastic. These structures are also known as 'saccharoidal. Quartzites and marbles are typical examples of this structure.

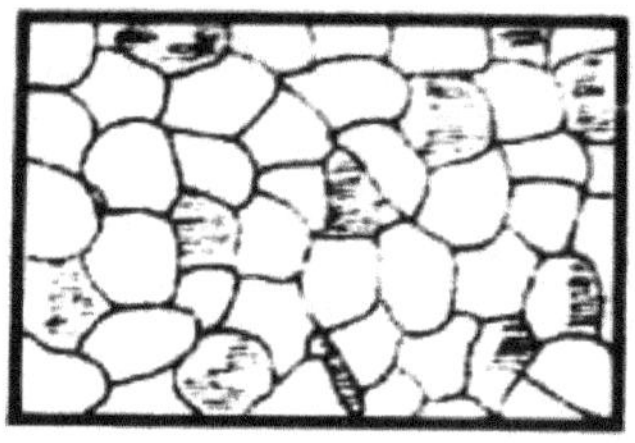

Granulose Structure (Marble)

5. Gneissose structure: It is a banded structure due to alteration of schistose (dark coloured) and granulose (light coloured) bands and is produced by highest grade of metamorphism, typically by regional metamorphism. The bands differ from ore another in colour, texture and mineral composition. Gneisses typically show this type of structure.

Gneissose Structure (Banded Gneiss)

Metamorphic Textures

The texture of a metamorphic rock is both a description of its constituent minerals along with their arrangement and size. There are two types of Textures; namely:

Foliated Textures

1. Slaty Texture– This texture is caused by the parallel orientation of microscopic grains. The name for the rock with this texture is slate, and the rock is characterized by a tendency to separate along parallel planes. This feature is a property known as *slaty cleavage.*

2. Phyllitic Texture– This texture is formed by the parallel arrangement of platy minerals, usually micas. The parallelism is often crenulated. The predominance

of micaceous minerals imparts sheen to the hand specimens. A rock with a phyllitic texture is called a phyllite.

3. Schistose Texture- This is a foliated texture resulting from the subparallel to parallel orientation of platy minerals such as chlorite or micas. Other common minerals present are quartz and amphiholes. A schistose texture lies between the parallel platy appearance of phyllite and the distinct banding of gneissic texture. The average grain size of the minerals is generally smaller than in gneiss. A rock with schistose texture is called schist.

4. Gneissic Texture- This is a coarsely foliated texture in which the minerals have been segregated into discontinuous hands, each of which is dominated by one or two minerals. These bands range in thickness from 1 mm to several centimetres. The individual mineral grains are macroscopic and impart a striped appearance to a hand specimen. Light-colored bands commonly contain quartz and feldspar and the dark bands are commonly composed of hornblende and biotite. A rock with a gneissic texture is called gneiss.

Non- Foliated Texture

1. Granoblastic Texture: The metamorphic rocks with no visible preferred orientation of mineral grains have a non-foliated granoblastic texture. Non-foliated rocks commonly contain equidimensional grains of a single mineral such as quartz, calcite, or dolomite.

Examples of such rocks are quartzite, formed from a quartz sandstone, and marble, formed from a limestone or dolomite and the third non-foliated rock is hornfels identified by its dense, fine grained, hard, blocky or splintery texture composed of several silicate minerals

Metamorphism

Metamorphism is a mineralogical and structural adjustment of solid rocks to physical and chemical conditions differing from those under which the rocks originally formed. Changes produced by surface conditions such as compaction are usually excluded. The most important agents of metamorphism include temperature, pressure, and fluids. Equally as significant are changes in chemical environment that result in two metamorphic processes:

1. Mechanical dislocation where a rock is deformed, especially as a consequence of differential stress; and

2. Chemical recrystallization where a mineral assemblage becomes out of equilibrium due to temperature and pressure changes and a new mineral assemblage forms.

Three types of metamorphism may occur depending on the relative effect of mechanical and chemical changes. Dynamic metamorphism, or cataclasis, results mainly from mechanical deformation with little long-term temperature change. Textures produced by such adjustments range from breccias composed of angular, shattered rock fragments to very fine-grained, granulated or powdered rocks with obvious foliation and lineation. Large, pre-existing mineral grains may be deformed as a result of stress. Contact metamorphism occurs primarily as a consequence of increases in temperature when differential stress is minor. A common phenomenon is the effect produced adjacent to igneous intrusions where several metamorphic zones represented by changing mineral assemblages reflect the temperature gradient from the high-temperature intrusion to the low-temperature host rocks; these zones are concentric to the intrusion. Because the volume affected is small, the pressure is near constant. Resulting rocks have equidimensional grains because of a lack of stress and are usually fine-grained due to the short duration of metamorphism. Regional metamorphism results from the general increase, usually correlated, of temperature and pressure over a large area. Grades or intensities of metamorphism are represented by different mineral assemblages that either give relative values of temperature or absolute values when calibrated against laboratory experiments. Regional metamorphism can be subdivided into different pressure-temperature conditions based on observed sequences of mineral assemblages. It may include an extreme condition, where partial melting occurs, called anatexis.

Other types of metamorphism can occur. They are retrograde metamorphism, the response of mineral assemblages to decreasing temperature and pressure; metasomatism, the metamorphism that includes the addition or subtraction of components from the original assemblage; poly-metamorphism, the effect of more than one metamorphic event; and hydrothermal metamorphism, the changes that occur in the presence of water at high temperature and pressure which affect the resulting mineralogy and rate of reaction.

Types of Metamorphism

Contact or Thermal Metamorphism

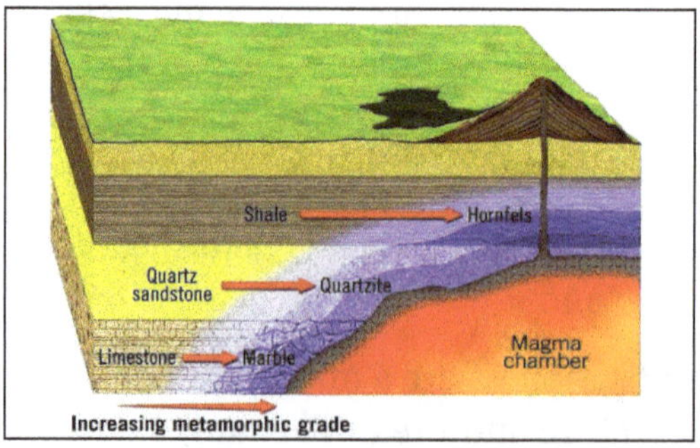

Contact metamorphism

Contact metamorphism is the name given to the changes that take place when magma is injected in the surrounding solid rock (country rock).

The changes that occur are greatest wherever the magma comes into contact with the rock because the temperatures are highest at this boundary and decrease with distance from it. Around the igneous rock that forms from the cooling magma is a metamorphosed zone called a contact metamorphism aureole.

Aureoles may show all degrees of metamorphism from the contact area to unmetamorphosed (unchanged) country rock some distance away. The formation of important ore minerals may occur by the process of metasomatism at or near the contact zone.

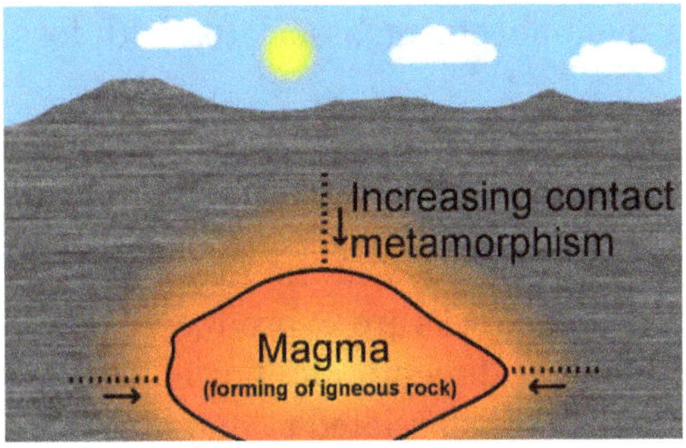

When a rock is contact altered by an igneous intrusion it very frequently becomes more indurated, and more coarsely crystalline. Many altered rocks of this type were formerly called hornstones, and the term hornfels is often used by geologists to signify those fine grained, compact, non-foliated products of contact metamorphism.

A shale may become a dark argillaceous hornfels, full of tiny plates of brownish biotite; a marl or impure limestone may change to a grey, yellow or greenish lime-silicate-hornfels or siliceous marble, tough and splintery, with abundant augite, garnet, wollastonite and other minerals in which calcite is an important component.

A diabase or andesite may become a diabase hornfels or andesite hornfels with development of new hornblende and biotite and a partial recrystallization of the original feldspar. Chert or flint may become a finely crystalline quartz rock; sandstones lose their clastic structure and are converted into a mosaic of small close-fitting grains of quartz in a metamorphic rock called quartzite.

If the rock was originally banded or foliated (as, for example, a laminated sandstone or a foliated calc-schist) this character may not be obliterated, and a banded hornfels is the product; fossils even may have their shapes preserved, though entirely recrystallized, and in many contact-altered lavas the vesicles are still visible, though their contents have usually entered into new combinations to form minerals that were not originally present.

The minute structures, however, disappear, often completely, if the thermal alteration is very profound. Thus small grains of quartz in shale are lost or blend with the surrounding particles of clay, and the fine ground-mass of lavas is entirely reconstructed.

Magmatic fluids coming from the intrusive rock may also take part in the metamorphic reactions. An extensive addition of magmatic fluids can significantly modify the chemistry of the affected rocks. In this case the metamorphism grades into metasomatism.

If the intruded rock is rich in carbonate the result is a skarn. Fluorine-rich magmatic waters which leave a cooling granite may often form greisens within and adjacent to the contact of the granite. Metasomatic altered aureoles can localize the deposition of metallic ore minerals and thus are of economic interest.

A special type of contact metamorphism, associated with fossil fuel fires, is known as pyrometamorphism.

Regional Metamorphism

Regional metamorphism is associated with the major events of Earth dynamics, and the vast majority of metamorphic rocks are so produced. They are the rocks involved in the cyclic processes of erosion, sedimentation, burial, metamorphism, and mountain building (orogeny), events that are all related to major convective processes in Earth's mantle.

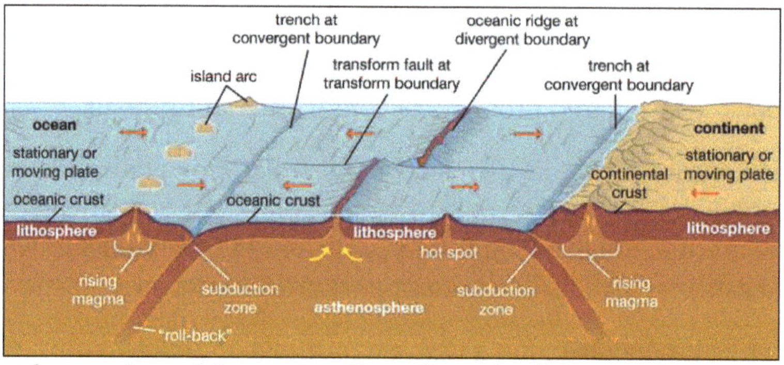

Crustal generation and destruction: Three-dimensional diagram showing crustal generation and destruction according to the theory of plate tectonics; included are the three kinds of plate boundaries—divergent, convergent (or collision), and strike-slip (or transform).

Most regionally metamorphosed rocks develop primarily in response to continent-continent collision and to collision between oceanic and continental plates. As a result, young metamorphic belts aligned roughly parallel to the present-day continental margins (e.g., the Pacific margin) as well as older metamorphic belts are used to infer the geometries of the continental margins at earlier periods in Earth history. Most of the world's mountain belts are at least partially composed of regionally metamorphosed rocks, with spectacular examples provided by the Alps, the Himalayas, the northern Appalachians, and the Highlands of Scotland. Although the processes that formed each of these mountain belts are broadly similar, in almost all such crustal events at different times and places, there is uniqueness as well as conformity to a general pattern. Metamorphic events in the Alps, the Urals, and the Himalayas all show specific differences: to unravel such differences and their significance is one of the major tasks of metamorphic petrology.

In areas of collision between oceanic and continental lithospheric plates such as the circum-Pacific region, the denser oceanic plate is subducted (carried into Earth's mantle) beneath the more buoyant continental lithosphere. Rapid subduction of the cool oceanic lithosphere perturbs the thermal regime in such a way that high pressures can be obtained at relatively low temperatures, thereby generating blueschists and eclogites (high-pressure facies series) from ocean-floor basalts transported down the subduction zone. Continued subduction of these rocks to great depth may eventually result in either (1) rising temperatures and partial melting of subducted rocks or (2) the melting of hydrated peridotite created by fluids released from metamorphic reactions in the subduction zone that rise into the overlying mantle wedge. These melts contribute to the formation of the volcanoes that overlie subduction zones in areas such as the Andes of South America, Japan, and the Aleutian Islands. Upward migration of subduction-related magmas also contributes to the development of paired metamorphic belts, in which high-pressure, low-temperature metamorphic rocks are flanked on the continental side by a parallel belt of low-pressure, high-temperature rocks. The latter rocks are thought to reflect perturbation of the crustal thermal regime by the passage of silicate melts generated above the subducting slab. Continued intrusion of magma over a period of time would cause an increase in crustal temperatures at relatively shallow depths and produce the high-temperature rocks adjacent to the high-pressure rocks generated in the subduction zone. Well-developed paired metamorphic belts are exposed in Japan, California, the Alps, and New Zealand.

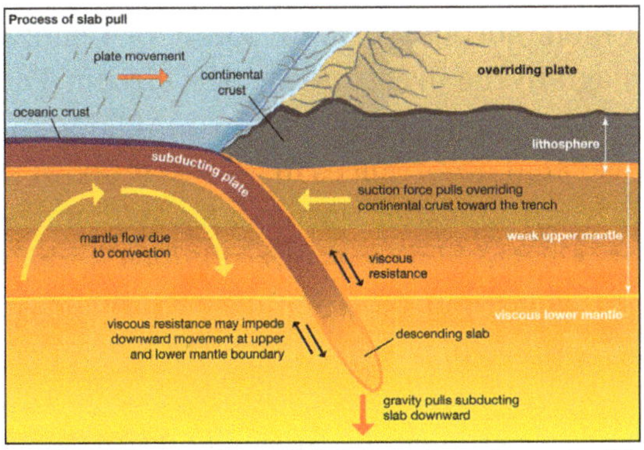

Slab pull process: The weight of the subducted slab may drag the rest of the tectonic plate toward the trench, a process known as slab pull, much as a tablecloth will pull itself off a table if more than half of the cloth is draped over the table's edge.

Data obtained from deep earthquakes in subduction zones indicate that a descending slab of oceanic lithosphere can remain intact to depths of several hundred kilometres before undergoing complete melting or fragmentation or both and being incorporated into the surrounding mantle. Clearly, the blueschists and eclogites exposed in orogenic belts around the world did not undergo such a process and were instead returned to Earth's surface. Most of the high-pressure rocks that have been studied from Japan, California, New Caledonia, the Alps, and Scandinavia record maximum pressures of 10–20 kilobars (about 9,900–19,700 standard atmospheres), corresponding to subduction to depths of approximately 35–70 km (about 22–44 miles). A few samples have been discovered in Norway, the Alps, and China that contain the mineral coesite, a high-pressure polymorph of quartz. Experimental studies on the stability of coesite imply minimum pressures of 30 kilobars (about 29,600 standard atmospheres) for these rocks, indicating burial or subduction to depths of approximately 100 km (62 miles). This is termed ultrahigh-pressure metamorphism (UHPM). These pressures are particularly noteworthy in that they are recorded in rocks derived from sedimentary rather than basaltic protoliths. Because of the low density, and hence greater buoyancy, of sediments relative to basalts, many geologists have argued that sediment subduction must be a rather limited process; the coesite-bearing metapelites (metamorphosed pelites) provide important evidence that sediment subduction can and does occur under certain circumstances.

The processes by which rocks that have been partially subducted are returned to the surface are not well understood. Models have been proposed to account for uplift and exposure of these high-pressure, high-density rocks; they include scraping material from the subducting plate against the overlying crustal lithosphere, upward flow of material in response to forced convection above the subducted slab, and removal of overlying thickened crust by low-angle extensional faulting. Testing these models requires considerable petrologic and structural work in areas where high-pressure rocks are exposed.

Most of the high-pressure rocks that are currently displayed in metamorphic belts around the world were metamorphosed in Mesozoic or Cenozoic time—that is, from some 252 million years ago to the present—e.g., the circum-Pacific belt, the Alps, the Greek Cyclades, and the Cordillera Betica in Spain. Older high-pressure rocks are known from only a few isolated occurrences in, for example, Wales, Bavaria, the ële de Groix off the coast of Brittany, and the Norwegian Caledonides (on the west coast of Norway). The general absence of high-pressure samples in the early rock record raises a number of interesting questions concerning Earth history. Some geologists have argued that the lack of well-developed high-pressure belts formed during Precambrian and Paleozoic time (4.6 billion to 252 million years ago) indicates that plate-tectonic processes have changed significantly throughout geologic time. Specifically, they claim that greater heat production in Archean time (about 4 billion to 2.5 billion years ago) would have produced hotter crustal geotherms, resulting in thin hot lithospheric plates whose mechanical behaviour may have been quite different from that of the present-day plates and hence may not have permitted formation of subduction zones. The increasing abundance of subduction-related metamorphic rocks with decreasing age in the rock record would thus reflect the gradual onset of plate tectonics as operative today. Others argue that the rock record is biased because of preferential erosion or thermal overprinting (development of a new mineralogy that may obliterate the original one) of old blueschists and eclogites. Thermal modeling studies suggest that blueschists will generally undergo heating and be converted to greenschist assemblages if exposure at Earth's surface does not occur within 100 million to 200 million years after high-pressure metamorphism. Early exposure at the surface also increases the chances for removal by erosion, however, resulting in a low probability for preserving blueschists greater than 100 million to 200 million years old. Geologists favouring generation of blueschists throughout Earth history but only selective preservation of these rocks also point to crustal rocks more than 2.5 billion years old that record metamorphism at depths of 25–40 km (15.5–24.8 miles). These medium-pressure facies series rocks imply that crustal thicknesses in early Earth were similar to those of the present day and thus that modern plate-tectonic processes may have operated from the early Precambrian to the present. This debate, though unresolved, emphasizes the substantial knowledge of the thermal structure of Earth and plate-tectonic processes that can be obtained from the study of metamorphic rocks.

Depending on the original geometry of Earth's lithospheric plates, subduction of oceanic crust beneath continental lithosphere may result in complete consumption of an ocean basin and subsequent collision between two continents. Collisions of this type have a long and complex history that may include initial formation of a paired metamorphic belt followed by extreme crustal thickening in response to the actual collision of the continents. The overthickened crust produced by the collision event will be gravitationally unstable and will undergo subsequent rapid erosion and possibly extensional faulting in order to return to a normal crustal thickness. Rocks metamorphosed in the early stages of collision may belong to a high-pressure facies series, reflecting the final stages of subduction of oceanic lithosphere, whereas the younger facies more typically belong to medium-pressure facies series. Metamorphic rocks exposed in former collision zones may thus have followed a variety of pressure-temperature-time paths, but paths showing rapid burial followed by heating and subsequent unroofing at moderate to high temperatures have been reported from many mountain belts around the world. Owing to the strong directed forces operative during collision, deformation typically accompanies metamorphism; rocks metamorphosed in response to continent-continent collision generally have fabrics showing a strong preferred orientation of mineral grains, folds on a variety of scales, and pre-, syn-, and postkinematic porphyroblasts. Examples of metamorphic belts produced in response to this type of collision include the Paleozoic Appalachian and Caledonides belts and the Mesozoic-Cenozoic Alpine and Himalayan belts.

Regionally metamorphosed rocks are also exposed in areas where the crust has been thinned by extensional faulting, such as the Basin and Range Province of the western United States. In this type of occurrence, areas of medium- and low-pressure facies series rocks that measure a few tens of kilometres in diameter are juxtaposed against unmetamorphosed sediments or very low-grade metamorphic rocks along low-angle extensional faults. (Metamorphic grades refer to the degree and intensity of the metamorphism: they are determined by the pressure and temperatures to which the rock has been subjected.) Such areas are generally referred to as metamorphic core complexes. Metamorphism in these complexes may or may not be related to the extensional event. In some instances, metamorphic rocks produced during much earlier events are simply unroofed and exposed by the faulting but show little or no recrystallization related to extension. In other cases, prolonged extension has resulted in an increased crustal geotherm, and relatively high-temperature metamorphism and magmatism is thus directly related to the extensional event. Immediately adjacent to the faults, the rocks may also be affected by dynamic metamorphism.

The facies associated with regional metamorphism include, at low grade, the zeolite and prehnite-pumpellyite facies. In areas belonging to high-pressure facies series, the rocks are predominantly in the blueschist and eclogite facies. Medium- and low-pressure facies series are typified by rocks belonging to the greenschist, amphibolite, and granulite facies.

Zeolite Facies

In the zeolite facies, sediments and volcanic debris show the first major response to burial. Reactions are often not complete, and typical metamorphic fabrics may be poorly developed or not developed at all. This is the facies of burial metamorphism.

The zeolite facies was first described from southern New Zealand, but similar rocks have now been described from many younger mountain regions of Earth, particularly around the Pacific margin and the European Alps. Typically, the rocks are best developed where reactive volcanic materials (often partly glassy) are common and the characteristic minerals include zeolites, which are low-density hydrated silicates, stable at temperatures rarely exceeding 300 °C (572 °F). Typical mineral assemblages include heulandite, analcite, quartz with complex clay minerals (montmorillonite), micaceous phases such as chlorite and celadonite, and the potassium feldspar adularia. At higher grades of metamorphism, the zeolite laumonite and the feldspar albite dominate the mineral assemblage. In New Zealand these are developed in a rock column that is about 15 km (9.3 miles) thick. Calcareous rocks (impure limestones) show very little response to this grade of metamorphism.

Prehnite-pumpellyite Facies

Along with the zeolite facies, the prehnite-pumpellyite facies received little attention until about 1950. The first rocks of the facies were described in New Zealand and Celebes. The facies is transitional, bridging the path to the blueschist facies or the greenschist facies. It is particularly well developed in graywacke-type sediments. The two minerals prehnite and pumpellyite replace the zeolite minerals of the zeolite facies and are themselves replaced by epidote minerals in the greenschist facies and by lawsonite and pyroxenes in the blueschist facies. Typical minerals in this facies are quartz, albite, prehnite, pumpellyite, chlorite, stilpnomelane, muscovite, and actinolite. Almost all the minerals are hydrated, and, except for chlorite, they bear little resemblance to the minerals of sediments. This facies has been most described from younger mountain ranges of the Pacific margin.

Blueschist Facies

Rocks of the blueschist facies represent deep metamorphism under conditions of a low thermal gradient. The characteristic locale for this type of metamorphism appears to be along a continental margin being underthrust by an oceanic plate. Regions in which blueschists are found are also regions of great seismic and volcanic activity, such as the Pacific margin. The best described examples of this class of metamorphism come from California, Japan, New Caledonia, Celebes, the Alps, and the Mediterranean region. At present there are no confirmed examples of glaucophane schists predating the Paleozoic Era. Because of the presence of the blue amphibole glaucophane and minerals such as garnet and jadeite, these schists are among the most attractive of metamorphic rocks.

Characteristic minerals of the facies include quartz, glaucophane, lawsonite, jadeite, omphacite, garnet, albite, chlorite, muscovite, paragonite, epidote, and kyanite. In calcareous rocks, calcite may be replaced by the high-pressure polymorph aragonite. In general, the facies is characterized by many high-density minerals reflecting a high pressure of formation.

Eclogite Facies

The eclogite facies was initially recognized in rocks only of basaltic composition, which are transformed at the pressure-temperature conditions of the eclogite facies into spectacular red and green rocks composed of the anhydrous mineral assemblage garnet plus omphacite. The garnet is rich in the high-pressure species pyrope, and the omphacite is rich in the high-pressure pyroxene jadeite. Small amounts of minerals such as kyanite, zoisite, and hornblende may be present. The rocks are of high density and frequently show little or no schistosity. It is now known that protoliths other than basalt also can be metamorphosed to pressures and temperatures characteristic of the eclogite facies, and a wide variety of mineral assemblages can be stable at these conditions, including several hydrous mineral phases. Minerals that have been observed in metapelites include magnesium-rich chloritoid and staurolite, kyanite, garnet, phengite (a muscovite mica with high magnesium and silicon and low aluminum content), chlorite, and talc. Experimental work shows that pelitic rocks composed primarily of talc and kyanite, which are referred to as whiteschists, can be stable from pressures of approximately 6 kilobars (about 5,900 standard atmospheres) up to greater than 30 kilobars (about 29,600 standard atmospheres). Minerals observed in eclogite-facies calcareous rocks include magnesite, dolomite, zoisite or epidote, and omphacite.

Because of the high density and composition, it was proposed long ago that part of the upper mantle might be made of eclogite. Such a view is supported by eclogitic intrusions in volcanic rocks and by eclogitic inclusions in diamond-bearing kimberlite, which must come from the upper mantle. Some workers also think that eclogites found in metamorphic terrains in Norway, California, U.S., and the European Alps could also come from the mantle by tectonic processes.

Early experimental work on eclogites of basaltic bulk composition suggested that eclogites could generally be stable only if water pressure was much lower than the lithostatic pressure (the stress exerted on a body of rock by surrounding rock), and the facies was thus thought to represent dry high-pressure metamorphism of basaltic protoliths. Subsequent work on the more diverse protolith compositions revealed, however, that a wide range of water pressures is possible in the eclogite facies and that fluid composition in equilibrium with the eclogite minerals also probably vary greatly. Indeed, fluid inclusions (tiny bubbles of fluid trapped within mineral grains) in eclogite samples provide evidence of fluids containing nitrogen, salts, and carbon dioxide in addition to water. Eclogite metamorphism is therefore not confined to dry environments but results instead from metamorphism of a variety of rock types at pressures above about 10 kilobars, corresponding to burial to approximately 35 km (21.7 miles), and at temperatures ranging from about 400 to 1,000 °C (752 to 1,832 °F). The temperatures of the eclogite facies overlap those of the greenschist, amphibolite, and granulite facies, but the higher pressures result in distinctly different mineral assemblages characterized by high-density mineral phases.

Greenschist Facies

The greenschist facies was once considered the first major facies of metamorphism proper. The name comes from the abundance of the green mineral chlorite in such rocks. Because chlorite and muscovite are ubiquitous and because both exhibit a platy crystal habit, these rocks normally show a highly developed foliation and often exhibit strong metamorphic differentiation. They have been described from practically every metamorphic terrain on Earth, from earliest Precambrian to the young mountain regions. In fact, many of Earth's oldest rocks (about 3 billion to 4.28 billion years old) of the continental shield areas are in this facies, classic examples of which are in the Appalachians, the Highlands of Scotland, New Zealand, the European Alps, Japan, and Norway.

The dominant minerals of greenschists formed from silicate-rich sediments include quartz, albite, muscovite, chlorite, epidote, calcite, actinolite, magnetite, biotite, and paragonite. Minerals less common include the manganese-rich garnet spessartine, stilpnomelane, kyanite, rutile, sphene (titanite), pyrophyllite, and chloritoid. Calcareous rocks are dominated by calcite, dolomite, and quartz; the major carbonate minerals are thermally stable. It is only when large quantities of water flush away carbon dioxide or keeps its partial pressure low that carbonate-silicate reactions take place and liberate carbon dioxide. The typical minerals of this facies have lower water contents than the zeolite facies minerals do.

Amphibolite Facies

The amphibolite facies is the common high-grade facies of regional metamorphism, and, like the greenschist facies, such rocks are present in all ages from all over the world. Their characteristic feature is the development of the most common amphibole, hornblende, in the presence of a plagioclase feldspar and garnet. The rocks are normally highly foliated or schistose. Many zones or isograds subdividing the facies have been recognized, and classic studies have been made in the Highlands of Scotland, New Hampshire and Vermont in the United States, Switzerland, and the Himalayas.

Characteristic minerals derived from pelitic rocks are quartz, muscovite, biotite, garnet, plagioclase, kyanite, sillimanite, staurolite, and orthoclase. Minerals derived from basaltic rocks include hornblende, plagioclase, garnet, epidote, and biotite. Those derived from calcareous rocks are calcite, diopside, grossular (garnet), zoisite, actinolite (hornblende), scapolite, and phlogopite. Minerals from magnesium-rich ultramafic rocks are chlorite, anthophyllite, and talc. In most common types, water is present in minerals only of the mica and amphibole families, and, with their water contents of only about 1 to 3 percent, dehydration is nearing its metamorphic climax.

Granulite Facies

In rocks of basaltic composition, the granulite facies is an anhydrous facies that results from progressive dehydration of amphibolites at high temperature. Rocks of other bulk compositions may retain some hydrous minerals, such as biotite and hornblende, but it is likely that water pressure is lower than lithostatic pressure during most granulite facies metamorphism. Evidence for relatively low water pressures comes from fluid inclusion data indicating carbon dioxide-rich fluid compositions and from preservation of some bulk compositions that should have undergone nearly total melting at granulite temperatures if water pressure had been equal to lithostatic pressure.

Rocks of this facies frequently have a granular texture quite similar to plutonic igneous rocks. Schistosity is only weakly developed. Typical minerals of the facies are quartz, alkali feldspar, garnet, plagioclase, cordierite, sillimanite, and orthopyroxene. In calcareous members, dolomite, calcite, diopside, and forsterite occur; and it is in this facies that minerals of the scapolite family are best developed. Small amounts of hornblende are often present. A rare mineral occurring in this facies is sapphirine. The rock type charnockite (from Tamil Nadu, India), essentially an orthopyroxene granite, is normally included in this facies.

It appears from experimental studies that during ultrametamorphism, when melting starts, the basic reactions which take place are of the type:

- Biotite + other minerals→ melt + residue

- Hornblende + other minerals→ melt + residue

The first melts to form are partly wet granitic or granodioritic melts, and phases such as biotite and hornblende break down by producing a partly wet melt from the least refractory phases in the rocks. They would persist to much higher temperatures in other systems of their own composition. The residue in the above equations is a granulite-facies metamorphic rock containing phases such as pyroxene and sillimanite. Thus it is probable, but certainly not universally accepted, that many granulites are formed only in the presence of a silicate liquid. This liquid may, of course, move to higher crustal levels.

Large areas of granulite facies rocks are confined almost entirely to Precambrian areas of the continents (those areas that were formed more than 541 million years ago), with well-developed areas exposed in Canada, India, Africa, Antarctica, Greenland, and the Adirondack Mountains of New York in the northeastern United States. Smaller areas of granulite facies rocks occur in younger mountain belts, with Paleozoic examples in New England (U.S.) and Brittany and Paleogene and Neogene examples (those formed between about 66 million and 2.6 million years ago) in British Columbia (Canada) and Timor. The apparent decrease in the volume of granulite facies rocks with decreasing age of metamorphism has led some geologists to postulate, that plate-tectonic processes might have changed significantly with time—specifically that steady-state continental geotherms were hotter in Precambrian time than at the present time. Some work on pressure-temperature-time paths in granulites also suggests that Precambrian granulites were metamorphosed along distinctly different paths than younger granulites, lending credence to models invoking changes in tectonic processes. An alternative hypothesis is that large volumes of granulites have been formed throughout Earth history but that they have not yet been exposed by erosion. Pressures calculated from fragments of granulite-facies metamorphic rocks carried to the surface in young volcanic eruptions suggest that the fragments were derived from the lower crust. It is likely that the lower crust is currently composed largely of granulite-facies rocks that may be exposed by future episodes of mountain building, but it is also possible that these granulites will prove to be different from their Precambrian counterparts. In order to resolve some of the controversies surrounding the origin and composition of granulites, it is necessary that considerable studies of these rocks be conducted in the future.

Hydrothermal Metamorphism

Hydrothermal metamorphism, also called metasomatism, may take place across wide regions of rock, thereby constituting a variant of regional metamorphism. It may alternately may take place in a limited, localized area and constitute a variant of local metamorphism.

Hydrothermal metamorphism takes place when hot, volatile solutions percolate into and react with the protolith, or the original rock. The heat of the intrusive igneous body and the hot volatile fluids serves to catalyze metamorphic reactions in the host rock. The incident fluids enhance ion mobility in the system and are highly reactive.

Both the intrusive magma and the associated volatile fluids may introduce elements which were not present in the protolith into the reaction process. The incident volatile fluids may also dissolve and remove elements originally abundant in the host rock. The chemical composition of the host rock may be extensively altered during the process of hydrothermal metamorphism. Hydrothermal metamorphism is therefore a type of allochemical metamorphism.

The hot, volatile fluids responsible for hydrothermal metamorphism may be derived from several sources. Such fluids may consist of volatiles escaping directly from the intrusive magma; they may also be composed of meteoric groundwater heated by the igneous intrusion or by the geothermal gradient.

High temperatures and pressures can expel water from hydrated minerals such as gypsum, talc, chalk, and clay. These sources of fluid are typically associated with local metamorphism. When an oceanic plate is subsumed at a boundary between two plates, water may be transported along with it and may subsequently take part in any metamorphic processes which occur. Seawater may also seep into the crust as two oceanic plates move apart. Such sources of volatile fluids are associated with regional metamorphism and the movement of crustal plates.

Much hydrothermal metamorphism occurs at the boundaries of oceanic plates. Plates which are moving apart allow seawater to percolate through the oceanic crust. As the seawater migrates, it heats and reacts with the host rock.

Large quantities of metals such as iron, cobalt, nickel, silver, gold, and copper are dissolved from the crust. When the hot, metal-laden fluid later comes into contact with cold seawater, sulfide and carbonate minerals precipitate to form deposits of metal ores. The copper ore which has been mined on Cyprus for several thousand years is thought to have been deposited in this way.

The presence of previously unavailable elements and the catalyzation provided by heat and volatile fluids enables a wide variety of mineral species to form during hydrothermal metamorphism. Many of these species contain elements not originally present in the host rock.

The process of hydrothermal metamorphism thus produces a greater variety of minerals than may be formed by thermal metamorphism alone. Silicates which are found in hydrothermal metamorphic deposits include quartz, garnet, wollastonite, olivine, topaz, and tourmaline. The halide fluorite may be present. Sulfide ores such as pyrite, chalcopyrite, sphalerite, and molybdenite and oxides such as magnetite, hematite, spinel, and corundum may also occur.

Local hydrothermal metamorphic deposits may grade into hydrothermal vein deposits. A vein forms when hot, mineral-bearing fluids deposit new materials along the walls of a preexisting fissure or crack in the host rock.

In the case where minerals merely exsolute from the aqueous solution to crystallize along the walls of the fissure, leaving the wall rock intact and unchanged, no metamorphism is involved. However, the minerals of the fissure walls often react with the passing volatile fluids. Some may be dissolved and transported away; some may recrystallize or crystallize as new mineral species. The formation of this type of vein deposit, called a hydrothermal replacement deposit, does constitute hydrothermal metamorphism.

Plutonic Metamorphism

This type of metamorphism is believed to occur at very great depths within the crust under conditions of very high pressure and very high temperature. The pressure prevailing is like hydrostatic pressure, i.e. the pressure intensity is the same in all the directions.

Such a pressure is sometimes called confining pressure. Due to such very high pressures, highly compact or dense varieties of minerals are formed. The minerals crystallizing in this metamorphic zone are likely to be stubbier or more equi dimensional than elongate. A striking characteristic of plutonic metamorphic rocks is their intimate association with intrusive igneous rocks.

Often these two types of rocks will alternate with each other in a single outcrop. For example, there may be a layer of granitic material, then of schist, then of granite, then of schist and so on.

An extreme example of plutonic metamorphism is the rock migmatite. In part these rocks have the banded or layered appearance of gneiss and yet in other parts of the outcrop, the constituent minerals will have the non-oriented, random, scattered pattern so typical of granite.

Dynamic Metamorphism

Dynamic (cataclastic) metamorphism is metamorphism of rock masses caused primarily by stresses that yield relatively high strain (deformation) rates. More simply, it is metamorphism resulting from deformation. The deformation may be dominantly brittle, in which case rock and mineral grains are broken and crushed, or it may be dominantly ductile, in which case plastic behavior and flow occur via structural changes within and between grains.' Temperatures during dynamic metamorphism are typically elevated and may be caused by the deformation process. Fluids commonly contribute to the metamorphic process, both by altering chemistry and by aiding recrystallization.

Both local and regional dynamic metamorphisms are recognized. At the local scale, in narrow zones from less than 1 cm to several meters wide, brittle or ductile deformation along faults and fold limbs causes rock to break, recrystallize, and even to melt. Similarly, both brittle and ductile deformation, as well as melting, occurs during impacts of extra-terrestrial bodies. Brittle and ductile deformation processes also operate at the regional scale.

The rocks produced at all scales by dynamic metamorphism are rocks composed of fragments of preexisting material (porphyroclasts), surrounded by a deformed matrix, the texture or mineral composition of which was produced by metamorphic processes. Such rocks, which fit into the broad category of clastic rocks, referred to as dynamoblastic rocks.

Occurrences of Dynoblastic Rocks

Faults are common within the crust of the Earth. Since faults are deformation zones, dynamoblastic rocks associated with faults are a common feature. In addition, folds and related deformation zones are relatively common in the roots of mountain belts. Even in zones in which newly formed rocks are only partially lithified, for example, in soft sediments on the seafloor, deformation may yield dynamically metamorphosed rocks. Particularly noteworthy among the local- to regional-scale zones of dynamoblastic rock are the mylonite zones associated with metamorphic core complexes and the melanges of outer metamorphic belts. Melanges are, in fact, mappable masses of dynamoblastic rock of local to regional dimensions. Impact structures with dynamoblasric rocks include Meteor Crater in Arizona and the Ries Basin of Germany.

Regional zones of dynamoblastic rocks occur at plate boundaries. Along spreading ridges, regional stress may be widespread enough to yield dynamically metamorphosed zones of rock. Perhaps more commonly, ductile deformation is concentrated in narrow zones within a regional terrane of schistose ultramafic rocks. Most local and regional zones of this type are probably subducted and are not preserved. Nevertheless, evidence of their existence is preserved locally in mantle slabs of accreted ophiolites. More commonly, oceanic crustal rocks are deformed along transform faults. Examples of rocks deformed in this way are exposed in the Sierra Nevada of California, in northern Italy, and on the island of Cyprus. Exposures of transform faults that transect the continents also reveal brittly and ductily deformed rocks, such as those along the San Andreas Fault System in California.

The most extensive development of dynamically metamorphosed rocks occurs in the mountain belts. Rocks of the transform fault zones may be accreted here, but most commonly, the regional zones of dynamoblastic rock are produced by deformation associated with the plate (and continent) collisions that yield the mountain range. At the shallower and cooler levels of orogens, melanges, formed by brittle deformation, ductile deformation, or both, are widespread. Well known examples include the melanges of the Franciscan Complex of California, and the Apennine Mountains of Italy."

Ductile deformation zones of regional extent are also common in the internal, high-temperature zones of the orogenic belts. Here, discrete fault lines are replaced by extensive zones of recrystallization and flow. Examples of such ductile deformation zones include some of the more regionally extensive mylonitic zones associated with metamorphic core complexes in the Rocky Mountain region, the Brevard Zone of the Southern Appalachian Orogen, faults in the Grenville Tectonic Zone in Ontario and the Moine Thrust of the Scottish Highlands.

Foliated Metamorphic Rocks

Gneiss

Gneiss is a foliated metamorphic rock identified by its bands and lenses of varying composition, while other bands contain granular minerals with an interlocking texture. Other bands contain platy or elongate minerals with evidence of preferred orientation. It is this banded appearance and texture - rather than composition - that define gneiss.

Gneiss: The specimen shown is about two inches (five centimeters) across. It is easy to see the "salt and pepper" banding of this rock.

Formation of Gneiss

Gneiss usually forms by regional metamorphism at convergent plate boundaries. It is a high-grade metamorphic rock in which mineral grains recrystallized under intense heat and pressure. This alteration increased the size of the mineral grains and segregated them into bands, a transformation which made the rock and its minerals more stable in their metamorphic environment.

Gneiss can form in several different ways. The most common path begins with shale, which is a sedimentary rock. Regional metamorphism can transform shale into slate, then phyllite, then schist, and finally into gneiss. During this transformation, clay particles in shale transform into micas and increase in size. Finally, the platy micas begin to recrystallize into granular minerals. The appearance of granular minerals is what marks the transition into gneiss.

Gneissic Granodiorite: An outcrop of gneissic granodiorite in the Zarembo Island area of southeastern Alaska.

Intense heat and pressure can also metamorphose granite into a banded rock known as "granite gneiss." This transformation is usually more of a structural change than a mineralogical transformation. Granite gneiss can also form through the metamorphism of sedimentary rocks. The end product of their metamorphism is a banded rock with a mineralogical composition like granite.

Composition and Texture of Gneiss

Although gneiss is not defined by its composition, most specimens have bands of feldspar and quartzgrains in an interlocking texture. These bands are usually light in color and alternate with bands of darker-colored minerals with platy or elongate habits. The dark minerals sometimes exhibit an orientation determined by the pressures of metamorphism.

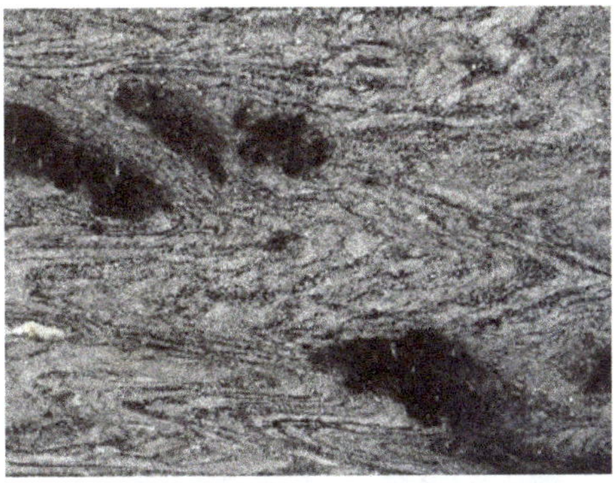

Folded Gneiss: A photograph of polished gneiss from the stock of a countertop vendor. The view shown in the photo is about 12 inches across.

Some specimens of gneiss contain distinctive minerals characteristic of the metamorphic environment. These minerals might include biotite, cordierite, sillimanite, kyanite, staurolite, andalusite, and garnet. Gneiss is sometimes named for these minerals, examples of which include "garnet gneiss" and "biotite gneiss."

Garnet Gneiss: A coarse-grained gneiss composed mainly of hornblende (black), plagioclase (white), and garnet (red) from Norway.

Garnet Gneiss: A cabochon cut and polished from garnet gneiss. A cabochon cut from this type of material is rarely seen, but it would be an interesting gem for a geologist. The stone is approximately 38 x 27 millimeters in size.

Uses of Gneiss

Gneiss usually does not split along planes of weakness like most other metamorphic rocks. This allows contractors to use gneiss as a crushed stone in road construction, building site preparation, and landscaping projects.

Some gneiss is durable enough to perform well as a dimension stone. These rocks are sawn or sheared into blocks and slabs used in a variety of building, paving, and curbing projects.

Some gneiss accepts a bright polish and is attractive enough for use as an architectural stone. Beautiful floor tiles, facing stone, stair treads, window sills, countertops, and cemetery monuments are often made from polished gneiss.

Garnet Gneiss: A photograph of polished garnet gneiss from the stock of a countertop vendor. The view shown in the photo is about 12 inches across.

Commercial Terminology

Don't be surprised if you see gneiss labeled as "granite" at a cabinet shop or monument company. In the dimension stone trade, any rock with visible, interlocking grains of feldspar is considered to be "granite" in that industry. Seeing gneiss, gabbro, labradorite, diorite, and other types of rock marketed as "granite" disturbs many geologists. However, this long-time practice of the dimension stone trade simplifies discussions with customers since not everyone knows the technical names of unusual igneous and metamorphic rocks.

Corundum Gneiss: This is a specimen of corundum gneiss from Gallatin Valley, Montana. This specimen is about four inches across and has a round blue sapphire crystal on the left side.

Gneiss in the Classroom

Small rock and mineral specimens about one inch in size are usually adequate for student examination and identification. However, many rock units, identified as gneiss in the field, have bands that are thicker than one inch. If samples of these rock units are broken into one-inch pieces, many of them will be too small to exhibit the banding features of gneiss. This will confuse many students and cause others to incorrectly identify the rock.

Teachers can avoid these problems by collecting specimens that clearly display a banded structure. Teachers who purchase specimens must examine them carefully before they are presented to students.

After students have learned to identify gneiss and many other rock types, presenting specimens of gneiss that do not exhibit banding can be a challenging way to have students: A) consider possibilities that are not obvious, and, B) realize that a single rock specimen may not adequately represent a rock unit.

Phyllite

Phyllite is a type of foliated metamorphic rock created from slate that is further metamorphosed so that very fine grained white mica achieves a preferred orientation. It is primarily composed of quartz, sericite mica, and chlorite.

Phyllite has fine-grained mica flakes in a preferred orientation, whereas slate has extremely fine clay flakes that achieve a preferred orientation, and schist has large flakes in a preferred orientation.Among foliated metamorphic rocks, it represents a gradation in the degree of metamorphism between slate and schist.

The minute crystals of graphite, sericite, or chlorite, or the translucent fine-grained white mica, impart a silky, sometimes golden sheen to the surfaces of cleavage, called "phyllitic luster".

The protolith (or parent rock) for phyllite is shale or pelite, or slate, which in turn came from a shale protolith. Its constituent platy minerals are larger than those in slate but are not visible with the naked eye. Phyllites are said to have a texture called "phyllitic sheen," and are usually classified as having formed through low-grade metamorphic conditions through regional metamorphism metamorphic facies.

Phyllite has good fissility (a tendency to split into sheets). Phyllites are usually black to gray or light greenish gray in color. The foliation is commonly crinkled or wavy in appearance.

Phyllite is commonly found in the Dalradian metasediments of northwest Arran. In north Cornwall, there are Tredorn phyllites and Woolgarden phyllites.

Rock or stone is a natural substance, a solid aggregate of one or more minerals or mineraloids. For example, granite, a common rock, is a combination of the minerals quartz, feldspar and biotite. The Earth's outer solid layer, the lithosphere, is made of rock.

Rock has been used by mankind throughout history. The minerals and metals found in rocks have been essential to human civilization.

Three major groups of rocks are defined: igneous, sedimentary, and metamorphic. The scientific study of rocks is called petrology, which is an essential component of geology.

Classification

At a granular level, rocks are composed of grains of minerals, which, in turn, are homogeneous solids formed from a chemical compound that is arranged in an orderly manner. The aggregate minerals forming the rock are held together by chemical bonds. The types and abundance of minerals in a rock are determined by the manner in which the rock was formed. Many rocks contain silica (SiO_2); a compound of silicon and oxygen that forms 74.3% of the Earth's crust. This material forms crystalswith other compounds in the rock. The proportion of silica in rocks and minerals is a major factor in determining their name and properties.

Rocks are geologically classified according to characteristics such as mineral and chemical composition, permeability, the texture of the constituent particles, and particle size. These physical properties are the end result of the processes that formed the rocks. Over the course of time, rocks can transform from one type into another, as described by the geological model called the rock cycle. These events produce three general classes of rock: igneous, sedimentary, and metamorphic.

The three classes of rocks are subdivided into many groups. However, there are no hard and fast boundaries between allied rocks. By increase or decrease in the proportions of their constituent minerals they pass by every gradation into one another, the distinctive structures also of one kind of rock may often be traced gradually merging into those of another. Hence the definitions adopted in establishing rock nomenclature merely correspond to more or less arbitrary selected points in a continuously graduated series.

Schist

Schist is a foliated metamorphic rock made up of plate-shaped mineral grains that are large enough to see with an unaided eye. It usually forms on a continental side of a convergent plate boundary where sedimentary rocks, such as shales and mudstones, have been subjected to compressive forces, heat, and chemical activity. This metamorphic environment is intense enough to convert the clay minerals of the sedimentary rocks into platy metamorphic minerals such as muscovite, biotite, and chlorite. To become schist, a shale must be metamorphosed in steps through slate and then through phyllite. If the schist is metamorphosed further, it might become a granular rock known as gneiss.

A rock does not need a specific mineral composition to be called "schist." It only needs to contain enough platy metamorphic minerals in alignment to exhibit distinct foliation. This texture allows the rock to be broken into thin slabs along the alignment direction of the platy mineral grains. This type of breakage is known as schistosity.

In rare cases the platy metamorphic minerals are not derived from the clay minerals of shale. The platy minerals can be graphite, talc, or hornblende from carbonaceous, basaltic, or other sources.

Muscovite schist: The dominant visible mineral in this schist is muscovite. Its platy grains are aligned in a common orientation, and that allows the rock to be split easily in the direction of the grain orientation. The specimen shown is about two inches (five centimeters) across.

Formation of Schist

Schist is a rock that has been exposed to a moderate level of heat and a moderate level of pressure. Let's trace its formation from its protoliths - the sedimentary rocks from which it forms. These are usually shales or mudstones.

In the convergent plate boundary environment, heat and chemical activity transform the clay minerals of shales and mudstones into platy mica minerals such as muscovite, biotite, and chlorite. The directed pressure pushes the transforming clay minerals from their random orientations into a common parallel alignment where the long axes of the platy minerals are oriented perpendicular to the direction of the compressive force. This transformation of minerals marks the point in the rock's history when it is no longer sedimentary but becomes the low-grade metamorphic rock known as "slate."

Chlorite schist: A schist with chlorite as the dominant visible mineral is known as a "chlorite schist." The specimen shown is about two inches (five centimeters) across.

Slate is has a dull luster, it can be split into thin sheets along the parallel mineral alignments, and the thin sheets will ring when they are dropped onto a hard surface. If the slate is exposed to additional metamorphism, the mica grains in the rock will begin to grow. The grains will elongate in a direction that is perpendicular to the direction of compressive force. This alignment and increase in mica grain size gives the rock a silky luster. At that point the rock can be called a "phyllite." When the platy mineral grains have grown large enough to be seen with the unaided eye, the rock can be called "schist." Additional heat, pressure, and chemical activity might convert the schist into a granular metamorphic rock known as "gneiss."

Types of Schist and their Composition

Garnet mica schist: This rock is composed of fine-grained muscovite mica with numerous visible grains of red garnet. The specimen shown is about two inches (five centimeters) across.

Mica minerals such as chlorite, muscovite, and biotite are the characteristic minerals of schist. These were formed through metamorphism of the clay minerals present in the protolith. Other common minerals in schist include quartz and feldspars that are inherited from the protolith. Micas, feldspars, and quartz usually account for most of the minerals present in schist.

Schists are often named according to the eye-visible minerals of metamorphic origin that are obvious and abundant when the rock is examined. Muscovite schist, biotite schist, and chlorite schist (often called "greenstone") are commonly used names. Other names based upon obvious metamorphic minerals are garnet schist, kyanite schist, staurolite schist, hornblende schist, and graphite schist.

Emeralds in mica schist: The large crystal is about 21 millimeters in length.

Some names used for schist often consist of three words, such as garnet graphite schist. In these cases the dominant metamorphic mineral's name is used second, and the less abundant mineral name is used first. Garnet graphite schist is schist that contains graphite as its dominant mineral, but abundant garnet is visible and present.

Garnet mica schist in thin section

In figure, this is a microscopic view of a garnet grain that has grown in schist. The large black grain is the garnet, the red elongate grains are mica flakes. The black, gray, and white grains are mostly silt or smaller size grains of quartz and feldspar. The garnet has grown by replacing, displacing, and including the mineral grains of the surrounding rock. You can see many of these grains as inclusions within the garnet. From this photo it is easy to understand why clean, gem-quality garnets with no inclusions are very hard to find. It is also hard to understand how garnet can grow into nice euhedral crystals under these conditions.

Schist as a Construction Material

Schist is not a rock with numerous industrial uses. Its abundant mica grains and its schistosity make it a rock of low physical strength, usually unsuitable for use as a construction aggregate, building stone, or decorative stone. The only exception is for its use as a fill when the physical properties of the material are not critical.

Schist as a Gem Material Host Rock

Schist is often the host rock for a variety of gemstones that form in metamorphic rocks. Gem-quality garnet, kyanite, tanzanite, emerald, andalusite, sphene, sapphire, ruby, scapolite, iolite, chrysoberyl and many other gem materials are found in schist.

Gem materials found in schist are often highly included. This is because their mineral crystals grow within the rock matrix, often including mineral grains of the host rock instead of replacing them or pushing them aside. The best metamorphic host rock for gem materials is usually limestone, which is easily dissolved or replaced when the gem materials are formed.

Slate

Slate is a metamorphic rock with a dull luster. The most common color of slate is gray, but it can also be brown, green, purple, or blue. Slate is formed when a sedimentary rock (shale, mudstone, or basalt) is compressed. Over time, slate may transition into other metamorphic rocks, such as phyllite or schist. You've likely encountered slate on a building or an old chalkboard.

Slate is the finest-grained metamorphic rock, which means you have to examine it closely to see its structure. It's also a foliated rock that displays what is called "slaty cleavage." Slaty cleavage occurs when fine clay flakes grow in a plane perpendicular to the compression. Striking slate along the foliation causes it to display fissility, breaking the rock into smooth, flat sheets.

Composition and Properties

Close examination of slate shows its cleavage and structure.

Slate is hard, brittle and crystalline. However, the grain structure is so fine that the crystals are not readily visible to the naked eye. When polished, slate appears dull, but is smooth to the touch.

Like many rocks, slate consists primarily of silicates, which are compounds made of silicon and oxygen. In slate, the elements mainly form the minerals quartz, muscovite (mica), and illite (clay, an aluminosilicate). Other minerals found in slate can include biotite, chlorite, hematite, pyrite, apatite, graphite, kaolinite, magnetite, feldspar, tourmaline, and zircon.

Some samples of slate appear spotted. These spots typically appear when iron is reduced. The spots may be spherical or appear as ovoids when stress deforms the rock.

Slate Localities

Penrhyn Slate Quarry, near Bethesda in north Wales, 1857.

In Europe, most slate is mined in Spain. It is also mined in the United Kingdom, and parts of France, Italy, and Portugal. Brazil is the second-biggest producer of slate. In the Americas, it's also found in Newfoundland, Pennsylvania, New York, Vermont, Maine, and Virginia. China, Australia, and the Arctic also have large reserves of slate.

Uses of Slate

Slate is the traditional material used to make chalkboards.

Most slate mined today is used to produce roofing tiles. Slate is a good material for this purpose because it does not absorb water, survives freezing and thawing well, and can be cut into sheets. For the same reason, slate is used for flooring, decorations, and paving.

Historically, slate has been used to produce writing tablets, whetstones, laboratory bench tops, whetstones, cemetery markers, and billiard tables. Because slate is an excellent electrical insulator, it was used for early electrical switch boxes. The Inuit used slate to make blades for ulus, a multi-purpose knife.

As shale is compressed, it becomes slate.

The word "slate" has held different meanings over the years and in various industries. In the past, the terms "slate" and "shale" have been used interchangeably. In the modern usage, geologists say shale is converted into slate. However, if you're looking at a partially metamorphosed rock, it's difficult to say whether it should be categorized as slate or as shale. One way to tell shale and slate apart is to strike it with a hammer. Slate emits a "tink" or a ring when struck. Shale and mudstone produce a dull thud.

A sheet of smooth stone used for writing might be referred to as a "slate," regardless of its composition. In addition to slate, writing boards have been made using soapstone or clay.

American coal miners may refer to the shale forming the floor and ceiling of a mine as slate. Fragments of shale separated from coal during processing may also be called slate. Although technically incorrect, the language is traditional.

Fossils in Slate

Ammonite Fossil in Slate.

Compared to other metamorphic rocks, slate forms under relatively low temperature and pressure. This makes it a good for fossil preservation. Even delicate structures may be preserved and readily discerned against the fine grain of the rock. However, the foliation pattern of slate can shear fossils or distort them when the rock cleaves.

Non-foliated Metamorphic Rocks

Hornfels

Hornfels is a fine-grained metamorphic rock that was subjected to the heat of contact metamorphism at a shallow depth. It was "baked" by heat conducted from a nearby magma chamber, sill, dike, or lava flow. Common temperatures for the formation of hornfels range from about 1300 to 1450 degrees Fahrenheit (700 to 800 degrees Celsius).

Hornfels: Hornfels is a fine-grained metamorphic rock without obvious foliation. It forms during contact metamorphism at shallow depth. The specimen shown is about two inches (five centimeters) across.

Because directed pressure does not play a significant role in the formation of hornfels, it is often made up of mineral grains that are equidimensional in shape and without a preferred orientation. The grain shape and orientation might also be inherited from its parent rock.

The name "hornfels" is assigned to a rock after considering its grain size, texture, and geologic history. As a result, hornfels does not have a specific chemical or mineralogical composition. It inherits its composition from the rocks that are metamorphosed plus the fluids involved in the metamorphic process. Interpreting composition, grain size, texture, and geologic history can make hornfels a very difficult rock to identify.

Parent Rocks and Protoliths

Hornfels is not a rock that is "deposited". Instead it is a rock type that forms when an existing rock is metamorphosed. The original rock that was metamorphosed is usually referred to as the "parent rock" or "protolith".

Banded Hornfels: Hornfels is often banded, especially when it forms from the metamorphism of sedimentary rocks. This specimen of hornfels is thought to have formed with sandstones and siltstones as the protoliths. The rock is about 6 inches (16 centimeters) across.

A variety of sedimentary, igneous, and metamorphic rocks can be the protolith of hornfels. Common protoliths of hornfels include sedimentary rocks such as shale, siltstone, sandstone, limestone and dolomite; igneous rocks such as basalt, gabbro, rhyolite, granite, andesite and diabase; or, metamorphic rocks such as schist and gneiss.

Characteristics of Hornfels

Hornfels often retains the stratification, large-scale geometry, and also some textural characteristics of the protolith. The changes of contact metamorphism that convert rocks to hornfels can include recrystallization, cementation, silicification, partial melting, and more.

The result is often a dense, hard, fine-grained rock that is generally homogenous and exhibits a semi-conchoidal fracture. Hornfels can be almost any color, but black, gray, brown, reddish and greenish rocks are common.

Hornfels Outcrop.

In figure, an outcrop of hornfels along the Dulles Greenway in Loudoun County, Virginia. These rocks were originally thin-bedded siltstones and sandstones. Then hot diabase intruded as sills above and below these rocks, metamorphosing them into hornfels. A normal fault, dipping to the lower right, offsets bedding and disrupts the fracture pattern.

On the basis of mineral composition, most occurrences of hornfels can be separated into one of three general groups:

- Pelitic Hornfels: usually derived from shale, slate, and schist.
- Carbonate Hornfels: usually derived from limestone, dolomite or marble.
- Mafic Hornfels: usually derived from mafic igneous rocks.

A wide range of minerals and mineral groups are encountered in hornfels. The minerals frequently seen include: actinolite andalusite, augite, biotite, calcite, chlorite, cordierite, diopside, epidote, feldspars, garnet, graphite, hornblende, kyanite, pyrite, scapolite, sillimanite, sphene, tourmaline, and vesuvianite.

Marble

Marble is a granular limestone or dolomite (i.e., rock composed of calcium-magnesium carbonate) that has been recrystallized under the influence of heat, pressure, and aqueous solutions. Commercially, it includes all decorative calcium-rich rocks that can be polished, as well as certain serpentines (verd antiques).

Petrographically marbles are massive rather than thin-layered and consist of a mosaic of calcite grains that rarely show any traces of crystalline form under the microscope. They are traversed by minute cracks that accord with the rhombohedral cleavage (planes of fracture that intersect to yield rhombic forms) of calcite. In the more severely deformed rocks, the grains show stripes and may be elongated in a particular direction or even crushed.

Marbles often occur interbedded with such metamorphic rocks as mica schists, phyllites, gneisses, and granulites and are most common in the older layers of Earth's crust that have been deeply buried in regions of extreme folding and igneous intrusion. The change from limestones rich in fossils into true marbles in such metamorphic regions is a common phenomenon; occasionally, as at Carrara, Italy, and at Bergen, Norway, recrystallization of the rock has not completely obliterated the organic structures.

Most of the white and gray marbles of Alabama, Georgia, and western New England, and that from Yule, Colorado, are recrystallized rocks, as are a number of Greek and Italian statuary marbles famous from antiquity, which are still quarried. These include the Parian marble, the Pentelic marble of Attica in which Phidias, Praxiteles, and other Greek sculptors executed their principal works, and the snow-white Carrara marble used by Michelangelo and Antonio Canova and favoured by modern sculptors. The exterior of the National Gallery of Art in Washington, D.C., is of Tennessee marble, and the Lincoln Memorial contains marbles from Yule, Colorado, Alabama (roof transparencies), and Georgia (Lincoln statue).

Even the purest of the metamorphic marbles, such as that from Carrara, contain some accessory minerals, which, in many cases, form a considerable proportion of the mass. The commonest are quartz in small rounded grains, scales of colourless or pale-yellow mica (muscovite and phlogopite), dark shining flakes of graphite, iron oxides, and small crystals of pyrite.

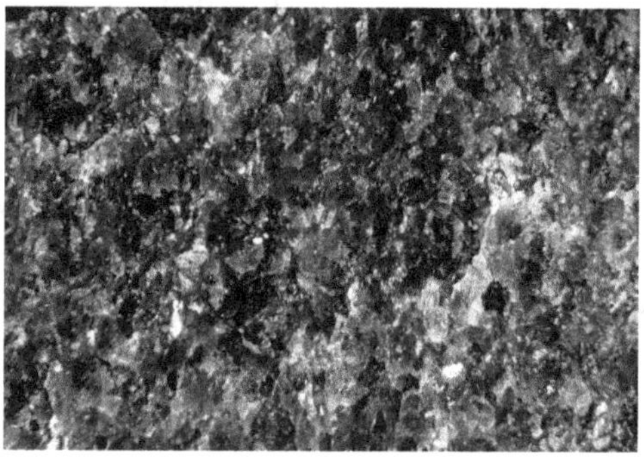

Close-up of metamorphic marble.

Many marbles contain other minerals that are usually silicates of lime or magnesia. Diopside is very frequent and may be white or pale green; white bladed tremolite and pale-green actinolite also occur; the feldspar encountered may be a potassium variety but is more commonly a plagioclase (sodium-rich to calcium-rich) such as albite, labradorite, or anorthite. Scapolite, various kinds of garnet, vesuvianite, spinel, forsterite, periclase, brucite, talc, zoisite, wollastonite, chlorite, tourmaline, epidote, chondrodite, biotite, titanite, and apatite are all possible accessory minerals. Pyrrhotite, sphalerite, and chalcopyrite also may be present in small amounts.

These minerals represent impurities in the original limestone, which reacted during metamorphism to form new compounds. The alumina represents an admixture of clay; the silicates derive their silica from quartz and from clay; the iron came from limonite, hematite, or pyrite in the original sedimentary rock. In some cases the original bedding of the calcareous sediments can be detected by mineral banding in the marble. The silicate minerals, if present in any considerable amount, may colour the marble; e.g., green in the case of green pyroxenes and amphiboles; brown in that of garnet and vesuvianite; and yellow in that of epidote, chondrodite, and titanite. Black and gray colours result from the presence of fine scales of graphite.

Bands of calc-silicate rock may alternate with bands of marble or form nodules and patches, sometimes producing interesting decorative effects, but these rocks are particularly difficult to finish because of the great difference in hardness between the silicates and carbonate minerals.

Later physical deformation and chemical decomposition of the metamorphic marbles often produces attractive coloured and variegated varieties. Decomposition yields hematite, brown limonite, pale-green talc, and, in particular, the green or yellow serpentine derived from forsterite and diopside, which is characteristic of the ophicalcites or verd antiques. Earth movements may shatter the rocks, producing fissures that are afterward filled with veins of calcite; in this way the beautiful brecciated, or veined, marbles are produced. Sometimes the broken fragments are rolled and rounded by the flow of marble under pressure.

The so-called onyx marbles consist of concentric zones of calcite or aragonite deposited from cold-water solutions in caves and crevices and around the exits of springs. They are not, in the strict sense, marble nor onyx, for true onyx is a banded chalcedony composed largely of silicon dioxide. Onyx marble was the "alabaster" of the ancients, but alabaster is now defined as gypsum, a calcium sulfate rock. These marbles are usually brown or yellow because of the presence of iron oxide. Well-known examples include the giallo antico ("antique yellow marble") of the Italian antiquaries, the reddish-mottled Siena marble from Tuscany, the large Mexican deposits at Tecali near Mexico City and at El Marmol, California, and the Algerian onyx marble used in the buildings of Carthage and Rome and rediscovered near Oued-Abdallah in 1849.

Unmetamorphosed limestones showing interesting colour contrasts or fossil remains are used extensively for architectural purposes. The Paleozoic rocks (from 251 million to 542 million years in age) of Great Britain, for example, include "madrepore marbles" rich in fossil corals and "encrinital marble" containing crinoid stem and arm plates with characteristic circular cross sections. The shelly limestones of the Purbeck Beds, England, and the Sussex marble, both of Mesozoic Era (from 251 million to 65.5 million years ago), consist of masses of shells of freshwater snails embedded in blue, gray, or greenish limestone. They were a favourite material of medieval architects and may be seen in Westminster Abbey and a number of English cathedrals. Black limestones containing bituminous matter, which commonly emit a fetid odour when struck, are widely used; the well-known petit granit of Belgium is a black marble containing crinoid stem plates, derived from fossil echinoderms (invertebrate marine animals).

Uses

Marbles are used principally for buildings and monuments, interior decoration, statuary, table tops, and novelties. Colour and appearance are their most important qualities. Resistance to abrasion, which is a function of cohesion between grains as well as the hardness of the component minerals, is important for floor and stair treads. The ability to transmit light is important for statuary marble, which achieves its lustre from light penetrating from about 12.7 to 38 mm (0.5 to 1.5 inches) from where it is reflected at the surfaces of deeper lying crystals. Brecciated, coloured marbles, onyx marble, and verd antique are used principally for interior decoration and for novelties. Statuary marble, the most valuable variety, must be pure white and of uniform grain size.

For endurance in exterior use, marble should be uniform and nonporous to prevent the entrance of water that might discolour the stone or cause disintegration by freezing. It also should be free from impurities such as pyrite that might lead to staining or weathering. Calcite marbles that are exposed to atmospheric moisture made acid by its contained carbon dioxide, sulfur dioxide, and other gases maintain a relatively smooth surface during weathering; but dolomite limestone may weather with an irregular, sandy surface from which the dolomite crystals stand out.

Marble statue of Julius Caesar.

The main mineral in marbles is calcite, and this mineral's variation in hardness, light transmission, and other properties in divers directions has many practical consequences in preparing some marbles. Calcite crystals are doubly refractive—they transmit light in two directions and more light in one direction; slabs prepared for uses in which translucency is significant are therefore cut parallel to that direction. Bending of marble slabs has been attributed to the directional thermal expansion of calcite crystals on heating.

Taj Mahal: marble portal
A marble portal of the Taj Mahal in Agra, India.

Quarrying

The use of explosives in the quarrying of marble is limited because of the danger of shattering the rock. Instead, channeling machines that utilize chisel-edged steel bars make cuts about 5 cm (2 inches) wide and a few metres deep. Wherever possible, advantage is taken of natural joints already present in the rock, and cuts are made in the direction of easiest splitting, which is a consequence of the parallel elongation of platy or fibrous minerals. The marble blocks outlined by joints and cuts are separated by driving wedges into drill holes. Mill sawing into slabs is done with sets of parallel iron blades that move back and forth and are fed by sand and water. The marble may be machined with lathes and carborundum wheels and is then polished with increasingly finer grades of abrasive. Even with the most careful quarrying and manufacturing methods, at least half of the total output of marble is waste. Some of this material is made into chips for terrazzo flooring and stucco wall finish. In various localities it is put to most of the major uses for which high-calcium limestone is suitable.

Marble quarry in Tuscany, Italy.

Novaculite

Novaculite: Specimen of novaculite showing its fine-grained texture and conchoidal fracture. Specimen is approximately 3 inches across.

Novaculite is a dense, hard, fine-grained siliceous rock that breaks with a conchoidal fracture. It forms from sediments deposited in marine environments where organisms such as diatoms (single-celled algae that secrete a hard shell composed of silicon dioxide) are abundant in the water. When the diatoms die, their silicon dioxide shells fall to the seafloor. In some areas these diatom shells are the primary ingredient of the seafloor sediments.

During diagenesis (the transformation from sediment to rock) the silicon dioxide from the diatom shells is transformed into chalcedony (a microcrystalline silicon dioxide). At this point the rock is chert. The chert is transformed into novaculite as further diagenesis and low-grade metamorphism recrystallize the chalcedony into microcrystalline quartz grains.

The two primary differences between chert and novaculite are:

1. Chert is composed mainly of chalcedony while novaculite is composed mainly of microcrystalline quartz grains;
2. Chert is a sedimentary rock, while novaculite is a chert that has experienced a higher level of diagenetic alteration and low-grade metamorphism.

Novaculite Localities

The most famous locality for novaculite is where the Arkansas Novaculite Formation outcrops in the Ouachita Mountains of central Arkansas and southeastern Oklahoma. It is a Devonian to Mississippian-age rock unit that ranges from about 60 feet thick in the northern Ouachitas to about 900 feet thick in the southern Ouachitas.

Outcrops of the Arkansas Novaculite Formation are prominent landscape features of the Ouachita Mountains. Compared to most other types of rock, novaculite is very resistant to chemical and physical weathering. This makes it a ridge-former and a cliff-former in the areas where it outcrops. Peaks, cliffs, and ridges formed by novaculite are prominent landscape features of the Ouachitas.

Arkansas novaculite sharpening stones

In figure, Sharpening stones made of Arkansas novaculite. The white stone has a coarse texture for initial sharpening, the mottled stone has an intermediate texture for resharpening, and the black stone has a very fine texture for honing an ultrasharp edge. The stones are used with a drop of oil that lubricates the sharpening strokes and keeps metal from loading the pore spaces in the stone. Stones are about two inches wide, six inches in length and 1/2 inch thick.

First use of Arkansas Novaculite

Novaculite ridges: Ridges of the Caballos Novaculite in the Lightning Hills of Brewster County, Texas. The Lightning overthrust crops out in the valley between the ridges. In the subsurface, the Caballos novaculite yields oil and gas from a tripolitic zone near the top of the rock unit and from fracture porosity in the lower portion of the rock unit.

Native Americans were the first people to mine the Arkansas Novaculite Formation. They noticed its conchoidal fracture and discovered that it could be knapped - just like flint - into projectile points, scrapers, and cutting tools. They mined novaculite, used it to manufacture cutting tools and weapons, and traded the material and products over a broad area. The Quapaw, Osage, Caddo, Tunica, Chickasaw, and Natchez tribes were especially involved in the mining.

Prehistoric people in other parts of the world have worked novaculite deposits to manufacture weapons and cutting tools. Novaculite and manufactured products from these areas were transported and traded across great distances.

A World-Famous Sharpening Stone

European settlers in the Ouachita region were the second people to mine the Arkansas Novaculite Formation. They valued it for a different reason. They found that novaculite could be used to sharpen metal tools and weapons. They soon began producing sharpening tools and trading them with distant partners.

Arkansas "whetstones," "oil stones," and "sharpening stones" became world-famous for their ability to produce a sharp edge on a metal blade. This created a demand for novaculite that was strong in the 1800s but declined as people used fewer blades that required resharpening. In the early 1900s, demand declined further as artificial abrasives and sharpening machines began to replace the sharpening stone. Although sharpening stones made with synthetic abrasives are cost-competitive with novaculite and perform well, a steady demand for novaculite still supports several producers of novaculite sharpening tools.

The Arkansas Novaculite Formation yields sharpening-grade stones in a range of textures. "Washita Stone" has the appearance of unglazed porcelain, a porosity of several percent and serves as a good stone for coarse sharpening. An extremely fine-grained material known as "Arkansas Stone" has almost no porosity and is an excellent tool for honing a razor-sharp blade. These stones are broken from the quarry with black-powder blasting, sawn to shape with a diamond saw, and then lapped to form a surface that is perfectly flat and smooth.

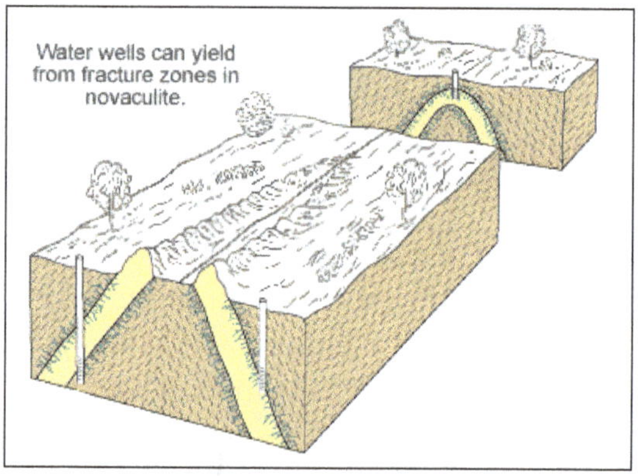

Water wells in novaculite: Novaculite is often a highly fractured rock unit that can serve as an adequate aquifer for private water supplies.

Other uses for Novaculite

Aggregate

Novaculite is a very durable rock that resists abrasion and is well suited for use as road base, railroad ballast, and rip-rap. Although it works especially well, its use in these applications is avoided. The reason: novaculite is so abrasive on metal that it causes excessive wear on excavating equipment used to mine it, on crushers and classifiers used to process it, and it wears out the beds of trucks that haul it. Novaculite is also not used as a concrete aggregate for the same reasons and because it reacts with cement to produce pop-outs (aggregate grains that separate from the concrete to produce a pit in the pavement surface).

Refractory

Novaculite's heat-resistant properties make it a good material for making refractory products. It has also been used in glass manufacturing, with some of it being used in Pyrex products. The abrasive properties of novaculite make it useful for manufacturing deburring media, files, and grinding media.

Tripoli

The upper portion of the Arkansas Novaculite Formation in some areas has a significant carbonate content. In these areas the novaculite weathers to yield a granular quartz residue with a very high silica content and a very fine grain size. This material is mined and processed into a product known as tripoli. Most of the tripoli used in the United States today is a filler or extender in plastics, rubber, paint, caulking compounds, and other products. Tripoli is added to soaps and scouring powders to serve as an abrasive. It is also used as an abrasive in metal finishing, woodworking, lapidary, and auto painting shops.

Reservoir Rock

Novaculite sometimes serves as a reservoir for oil and natural gas. Several oil and gas fields in the Ouachita overthrust belt of Oklahoma and Texas produce from the Caballos Novaculite. Tripolitic chert zones near the top of the rock unit can have significant porosity, and the highly fractured novaculite is another form of porosity. Areas underlain by fractured novaculite are also preferred drilling sites when drilling for groundwater.

Gold Testing

Small blocks of black novaculite are also used in the "acid test" for determining the gold content of jewelry. In this, the jeweler rubs the suspected gold item across a fine-grained block of black novaculite to produce a tiny streak of metal. A drop of aqua regia (a mixture of hydrochloric acid and nitric acid) of known concentration is placed on the streak. If the streak disappears, it was dissolved by the aqua regia. Aqua regia solutions of different concentration will dissolve different karat weights of gold. Standard aqua regia solutions have been developed to identify gold of 10k, 12k, 14k, 18k, 20k, and 22k purity.

Quartzite

Quartzite is a nonfoliated metamorphic rock that consists mostly of quartz. It's usually a white to pale gray rock, but occurs in other colors, including red and pink (from iron oxide), yellow, blue, green, and orange. The rock has a grainy surface with a sandpaper texture, but polishes to a glassy shine.

Formation of Quartzite

Quartzite forms when pure or nearly pure quartz sandstone undergoes heating and pressure. Usually this is caused by tectonic compression. The sand grains of sandstone melt and recrystallize, cemented together by silica.

Quartzite arenite is the intermediate stage between sandstone and quartzite. Arenite is still considered to be a sedimentary rock, but it has extremely high quartz content.

However, it's difficult to identify the transition from sandstone to quartzite. Some geologists use the term "quartzite" to refer to metamorphic rocks consisting almost exclusively of quartz. Here, quartzite is identified by the way it fractures across grain boundaries, while arenite breaks around them. Other geologists simply identify "quartzite" as a tightly-cemented rock found above or below a band of sedimentary quartz rock.

Quartzite Composition

Quartzite consists almost entirely of silicon dioxide, SiO_2. If the purity is about 99% SiO_2, the rock is called orthquartzite. Otherwise, quartzite commonly contains iron oxide and may contain trace amounts of the minerals rutile, zircon, and magnetite. Quartzite may contain fossils.

Properties

Quartzite has a Mohs hardness of 7, which is comparable to that of quartz and considerably harder than sandstone. Like glass and obsidian, it breaks with a conchoidal fracture. Its coarse texture makes it difficult to hone to a fine edge. Under magnification, quartzite's interlocking crystal structure becomes apparent.

Magnified thin section of quartzite displays its mineralogy.

Quartzite Localities

Quartzite cliffs surround Lake Oberon in Tasmania, Australia.

Quartzite forms at convergent tectonic plate boundaries. Converging plates bury sandstone and exert compression. As the boundary folds, mountains arise. Thus, quartzite is found in folded mountain ranges worldwide. While erosion weathers softer rock away, quartzite remains, forming peaks and cliffs. The rock also litters mountain sides as scree.

In the United States, you can find quartzite in eastern South Dakota, southwestern Minnesota, the Wasatch Range of Utah, the Baraboo Range of Wisconsin, Central Texas, near Washington, D.C., portions of Pennsylvania, and the mountains of Arizona and California. The town of Quartzite in Arizona takes its name from the rock in the nearby mountains.

Quartzite occurs throughout the United Kingdom, the La Cloche Mountains in Canada, and the Rhenish Massif in Continental Europe, Brazil, Poland, and the Chimanimani Plateau of Mozambique.

Uses

Quartzite's strength and toughness lends itself to many uses. Crushed quartzite is used in road construction and for railway ballast. It is used to make roofing tiles, stairs, and flooring. When cut and polished, the rock is quite beautiful, as well as durable. It is used to make kitchen countertops and decorative walls. High-purity quartzite is used to make silica sand, ferrosilicon, silicon carbide, and silicon. Paleolithic humans sometimes made stone tools out of quartzite, although it was harder to work than flint or obsidian.

Quartzite versus Quartz and Marble

Quartzite is a metamorphic rock, while quartz is an igneous rock that crystallizes from magma or precipiates around hydrothermal vents. Sandstone under pressure becomes quartz arenite and quartzite, but quartzite does not become quartz. The construction industry further complicates the matter. If you buy "quartz" for countertops, it is actually an engineered material made from crushed quartz, resin, and pigments and not the natural rock.

Another rock commonly confused with quartzite is marble. Both quartzite and marble tend to be pale-colored, non-foliated rock. Despite having a similar appearance, marble is a metamorphic rock made from recrystallized carbonate minerals, not silicates. Marble is softer than quartzite. An excellent test to distinguish the two is to apply a bit of vinegar or lemon juice to the rock. Quartzite is impervious to weak acid etching, but marble will bubble and retain a mark.

References

- Metamorphic-rock, science: britannica.com, Retrieved 25 February, 2019
- Metamorphic-textures-and-structures: geologyuniverse.com, Retrieved 6 July, 2019

- Metamorphism, science: britannica.com, Retrieved 11 April, 2019
- Contact-metamorphism: geologyin.com, Retrieved 15 August, 2019
- Regional-metamorphism, metamorphic-rock, science: britannica.com, Retrieved 7 May, 2019
- Hydrothermal-metamorphism: geologyin.com, Retrieved 13 January, 2019
- Types-of-metamorphism-geology, geology: yourarticlelibrary.com, Retrieved 21 June, 2019
- Dynamic, metamorphic, rocks: chemeketa.edu, Retrieved 11 March, 2019
- Gneiss, rocks: geology.com, Retrieved 22 May, 2019
- Phyllite: worldofstones.com, Retrieved 24 April, 2019
- Schist, rocks: geology.com, Retrieved 12 June, 2019
- Slate-rock-definition-composition-and-uses: thoughtco.com, Retrieved 16 May, 2019
- Hornfels, rocks: geology.com, Retrieved 6 February, 2019
- Marble-rock, science: britannica.com, Retrieved 26 August, 2019
- Novaculite, rocks: geology.com, Retrieved 1 July, 2019
- Quartzite-rock-geology-and-uses: thoughtco.com, Retrieved 31 January, 2019

5

Understanding Lithology

The description of a rock unit in terms of its physical characteristics visible at outcrop, in hand or core samples, or by using low magnification microscopy is known as lithology. The topics elaborated in this chapter will help in gaining a better perspective about the discipline of lithology as well as related disciplines of lithostratigraphy and stratigraphy.

The identification of a bed's lithology is fundamental to all reservoir characterization because the physical and chemical properties of the rock that holds hydrocarbons and water affect the response of every tool used to measure formation properties. Understanding reservoir lithology is the foundation from which all other petrophysical calculations are made. To make accurate petrophysical calculations of porosity, water saturation (S_w), and permeability, the various lithologies of the reservoir interval must be identified and their implications understood. Lithology means "the composition or type of rock such as sandstone or limestone." These few words belie a host of details about reservoir rocks:

- Their depositional and diagenetic history.
- Pore structure.
- Mineralogy.

The term lithology is used as a gross identification for a rock layer in the subsurface and uses familiar names such as:

- Sandstone (or sand).
- Limestone.
- Dolostone (or dolomite).
- Claystone (or clay).
- Chert.
- Coal.
- Shale (or mudrock).
- Diatomite.
- Halite.

- Anhydrite.
- Gypsum.
- Tuff.

The term rock type is a more detailed description than lithology because it reflects the natural groupings of pore systems that produce recognizable properties used to predict:

- Flow properties.
- Volumes.
- Fluid saturations.

Lithology focuses on grains, while rock type focuses on pores. The list of rock types contains more than 250 classifications. Another term used in the literature is the Greek equivalent "petrofacies." (Another term that is finally dropping from use is the incorrectly named "electrofacies.")

Direct Determination

Obtaining a physical sample of the reservoir is the surest way to unambiguously determine lithology and rock type, but obtaining this physical sample is not always easy. Mud logs are the first choice in wildcat wells, but exact assignment of a rock fragment to a particular depth is not without error. The size of the individual rock sample examined at the surface is rather small because it is limited by the size of drill cuttings and rock strength. Weak rocks, ones without cement, are often reduced to original detrital grain size by the drilling process, making it difficult to determine rock type, but still possible to determine lithology. Once the well is drilled and logged and rock layers are marked for further study, rock samples can be obtained through the use of wireline core takers or sidewall core drills.

Recovery of sidewall samples is not always a sure thing, and we often fail to get rock samples from zones of interest. In a wildcat well with a zone of high interest identified with logs, rock samples, and fluid samples, we can plug back the well several hundred feet, set a whipstock, perform a parallel sidetrack (called a bypass hole), and then take a whole core across the zone of interest. Because the two holes are parallel, we know exactly at which depth to swap out the drill bit with a core barrel. The resulting whole core can be sampled, and sophisticated core analysis can be used to identify the rock type of each zone and determine petrophysical parameters, which are used to refine the formation evaluation from the log data.

Indirect Determination

If no direct rock sample is available in a given zone, log responses must be used alone to simultaneously determine lithology, porosity, and fluid saturations. Rock typing is much more involved and requires the use of commercially available catalogs of analog dataor locally collected data to pinpoint petrophysical properties that can be used to refine porosity and fluid-saturation calculations. If, in a given wellbore, one is lucky enough to encounter parts of the desired reservoir below the free-water level, then fluid saturation is known and gross lithology can be determined from routine wireline logs such as density, neutron, and acoustic tools. All wireline service companies provide charts and answer products that convert wireline logging data into lithology; however, few beds are simple homogeneous layers. If some of the layers are beneath the resolving thickness of the tool, some average response of the layers, depending on bed and borehole geometry, can be observed. Single layers of salt (NaCl), anhydrite, gypsum, coal, or tuft are easily identified when several feet thick or thicker; however, salt-plugged sandstone can mimic gas-filled clean sandstone, or a very shaly rock type can hide the presence of gas.

Mica or potassium feldspar, when abundant in sandstones, can confuse lithology determination on the basis of the gamma ray (GR) response. The more sophisticated GR spectroscopy tools can help to identify the chemical species in a given rock and often can lead to deducing its lithology. However, no single wireline log or set of logs can determine pore size distribution and pore throat size distribution. These data are used to establish rock type, resulting fluid flow characteristics, and other petrophysical parameters needed to determine fluid saturation. If every reservoir in our wildcat well is hydrocarbon filled, the problem of quantification becomes more difficult, particularly when formation water resistivity, R_w, is unknown. In difficult areas, a follow-up well and a whole core, drilled with oil-based mud that is taken high in the hydrocarbon column, must be drilled to guarantee no in-situ water displacement. Subsequent sophisticated core analysis can determine the value of in-situ water salinity. If coring is not an option, a well will have to make a significant amount of water to test. In pressure-depletion reservoirs that never produce much water, the accurate determination of water salinity may never happen if the reservoir is drilled with oil-based mud, negating the use of the spontaneous potentials (SP) log.

Use of Acoustic Logs

Acoustic velocity is primarily a function of the rock matrix and can be used to identify different lithologies and for stratigraphic correlations. A variety of crossplot techniques, using acoustic measurements alone, or in combination with other porosity logs (neutron and density), have been devised to assist in lithologic identification. In particular, the M-N and mineral-identification-plot (MID) techniques use all three porosity logs in different combinations. Before lithology determination, the individual log measurements must be corrected for influences of:

- Gas effect.
- Secondary porosity.

- Bad hole conditions.
- Shaliness.

In general, the MID plot is more sensitivity to lithology, gas, and secondary porosity and provides superior results to M-N plots. Crossplots, using a variety of log measurements or combinations of measurements can be used to resolve specific lithologic problems related to local or regional geology.

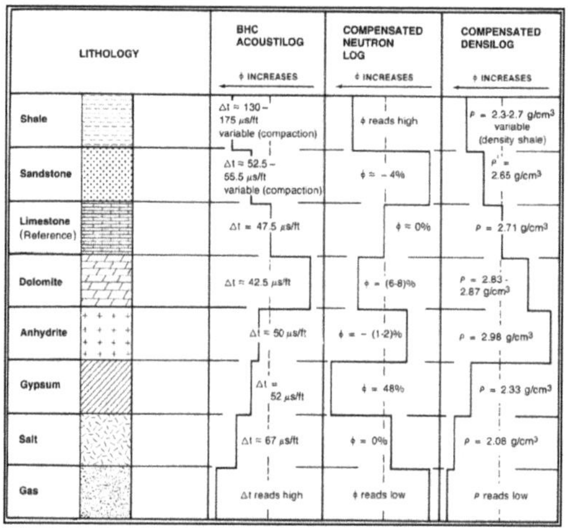

Figure: Generalized comparison of acoustic-, neutron-, and density-log responses in common reservoir lithologies and to gas.

The ratio of compressional to shear velocity, V_p/V_s, is an effective lithology indicator because each lithology exhibits a defined trend that is independent of porosity and depth. However, because the V_p/V_s ratio is affected by formation anisotropy, the ratio values may not be absolute indicators of a particular lithology.

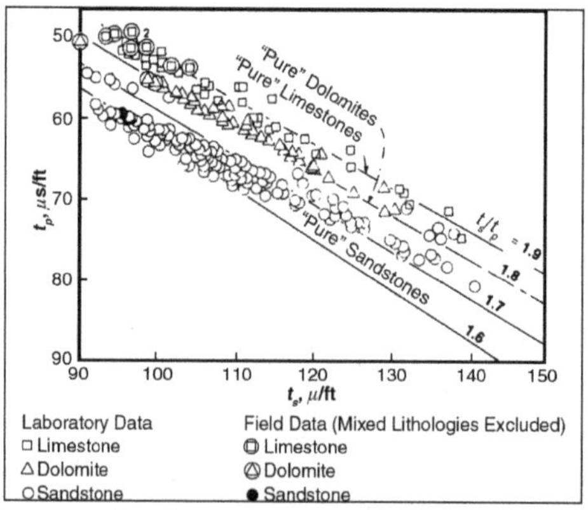

Figure: **Δt_p vs. Δt_s** crossplot demonstrates sensitivity to variations in lithology.

The addition of shear slowness to lithology identification provides a more robust result that can be particularly useful in cased-hole evaluations where density logs are not available. In figure, the combined use of shear slowness and cased-hole neutron porosity results in enhanced-porosity determination in a complex lithology. Crossplots of V_p/V_s ratio vs. compressional travel time, Δt_c, facilitate identification of lithology trends with respect to porosity and lithology. Recent studies of complex carbonate reservoirs indicate that the V_p/V_s ratio is a function of porosity and in some cases, can differentiate higher-permeability facies.

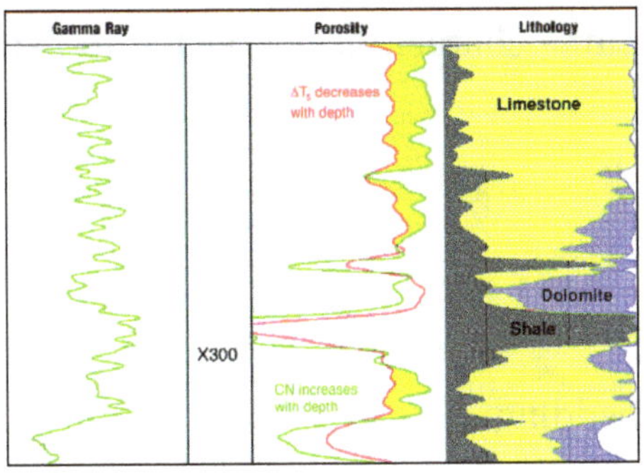

Shear slowness as a lithology indicator. The cased-hole apparent compensated-neutron porosity (Track 2) appears to increase with depth, while the dipole shear-velocity data decrease with depth. This indicates a lithology change rather than a porosity change, which is confirmed in the lithology track.

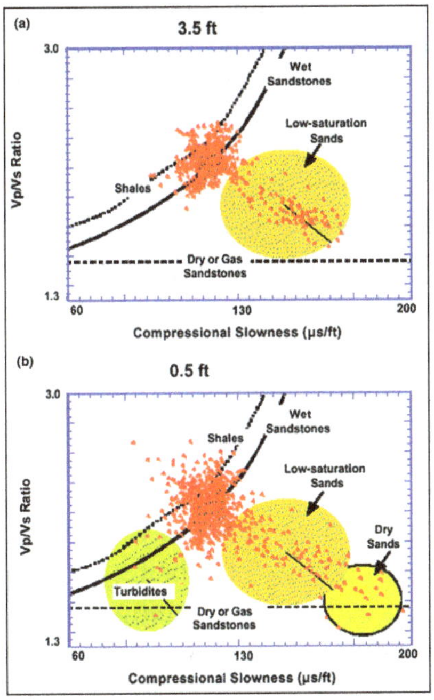

Figure: V_p/V_s vs. Δt_p crossplot. In addition to the low-saturation sands identified in (a), the high-resolution result (b) also resolves turbidite and dry sand intervals.

Determining Lithology

Geologists are trained to describe rocks, based on:

- Outcrops

- Cuttings
- Cores
- More-detailed mineralogical measurements

They identify certain log-curve characteristics related to particular depositional environments; for example:

- Coarsening upwards
- Fining upwards
- Massive bedding
- The scales of interbedding

For reservoir petrophysical evaluations, a geologist must be on the technical team, and there must be cooperation between the geologists and engineers.

The lithology of a new oil or gas reservoir is understood on a preliminary basis by the wellsite geologist's description of mud-shaker cuttings and possibly by a few cores that are cut if the reservoir interval is sufficiently long. Lithology is also determined from the logs, because each main reservoir lithology has characteristic responses. Frequently, lithologies are derived by pattern recognition of the GR-, density-, and neutron-log responses.

Some subsequent delineation wells are likely to be cored over the entire of the reservoir interval. The geologist will make detailed descriptions of these cores and order a number of thin-section, scanning electron microscope (SEM), X-ray diffraction (XRD), and mercury-injection capillary pressure (MICP) or pore-size measurements on various rock types that have been identified. From these data and the routine-core-analysis data, geologists can construct their interpretation of the reservoir's depositional environment and insights into the nature of its pore system and mineralogy. Geologists typically prepare reservoir cross sections with seismic traces, well logs, and core descriptions to illustrate the depositional environments, rock types, and internal geometries.

The lithology of a reservoir impacts the petrophysical calculations in numerous ways. The depositional environment and sediments being deposited will define the grain size, its sorting, and its distribution within the reservoir interval. In most sandstone reservoirs, the depositional environment controls the porosity/permeability relationship.

The mineralogy of a reservoir results from a combination of its depositional and diagenetic histories. For a sandstone reservoir, the depositional environment controls the percentages of:

- Quartz
- Chert

- Feldspar
- Detrital clay-mineral grains and the other matrix material

These materials must be measured and their variations within the reservoir interval quantified. The diagenetic history determines the extent to which:

- Portions of the grains have been leached away.
- Cements such as calcite, siderite, or pyrite have been deposited.
- Authigenetic clay minerals formed.

The diagenetic history can be complicated and can be impacted by differences in burial history from one part of a reservoir to other parts, or by aspects of the hydrocarbon-filling history.

For carbonate reservoirs, most of the same factors as discussed for sandstones come into play, but the mineralogical considerations are different. For carbonate formations, the rock formations typically consist of interbedded sequences of carbonates, dolomites, anhydrite, salt, and shale layers. The keys to reservoir development within the carbonate layers are the original grain size and how it has been altered by chemical diagenetic processes. As these chemical reactions take place, the pore-size distribution and porosity level will change (e.g., by dolomitization). Carbonate-reservoir porosity is also greatly enhanced by weathering, dissolution, and fracturing.

As well as the basic mineralogy, clay-mineral properties are also of particular importance to the petrophysical calculations in the reservoir intervals. There are many types of clay minerals, and their impacts on well logs are quite different. Particularly, there may be differences between clay minerals that form shale layers and claystones and those that occur within sandstone intervals. While the detrital clay minerals in sandstones will likely be the same as the clay minerals in the shales, the authigenetic clay minerals in the sandstone pore system can be quite different. The types and amounts of the various clay minerals impact the rock pore system by affecting its electrical properties and permeability characteristics. A few of the aspects of mineralogy that impact the petrophysical calculations are the extent to which:

- Various heavy minerals impact the density log, for example:
    - Anhydrite
    - Calcite
    - Dolomite
    - Granites
    - Pyrite
    - Siderite

- Various light minerals impact the density, neutron, and sonic logs, like:
    - Coal
    - Halite salt
- Various radioactive minerals impact the GR log, such as:
    - Uranium
    - Thorium
    - Potassium salts such as K-feldspar
- Various electrically conductive minerals impact the resistivity logs, like:
    - Clay minerals
    - Pyrite

Clay-mineral Properties

Clay minerals are, in general, composed of layered alumina and silicate molecules, and the properties of the various clay minerals vary widely. Some swell when wet, are plastic, and can easily deform, while others are hard and dense. Clay minerals are extremely fine-grained, and those with the smallest grain size have a very high surface-area-to-volume ratio. Clay minerals (e.g., chlorite, illite, kaolinite, smectites, and mixed-layer clays) generally impair the permeability and porosity of the pores in which they reside; the permeability is sometimes impacted by an order of magnitude or more. However, it is the smectites (one of which is montmorillonite) that often cause very significant effects on the petrophysical measurements of porosity and water saturation.

In smectites, exchange cations and adsorbed water molecules are loosely bound between the silicate layers. Dehydration occurs whenever there is low humidity or an elevated temperature (e.g., in a dried-up lakebed or in a brick kiln). Loss of this adsorbed water is even more rapid at temperatures above the boiling point of water. For smectites, this is a problem during core analysis because the extraction and drying of the core samples is an essential step in the measurement procedures.

For resistivity logging, smectite clay minerals pose a further problem. The exchange cations and adsorbed water molecules lead to smectite exhibiting excess electrical conductivity. This occurs as exchange cations (e.g., sodium, calcium) migrate from site to site on the clay surfaces. The clays thus exhibit a lower resistivity and, in most cases, depress the bulk resistivity of reservoir rocks in which they reside. Cation exchange also occurs with the other clay minerals, but to a lesser extent.

Rock formations of pure clay minerals are rare. More typically, several species of clay mineral are associated together with clay-sized and silt-sized quartz, mica, and other rock grains. This association is known widely as "shale." In sandstone reservoirs, clay-mineral content typically ranges from 0 to approximately 10% BV. In shales, clay minerals occupy approximately 20 to 40% BV, and the remainder is often very-fine-grained quartz, volcanic minerals, carbonates, and organic matter. Besides the adsorbed water on the clay minerals, shales and authigenetic clay minerals also include additional formation water held in their micron-sized pore system by capillary retention. This water cannot be produced from the formation and is referred to as “capillary water.”

During the burial of the sediments over geological time, the overburden stress and the pore-fluid pressures increase. The net result is that water is expelled from the shale beds into surrounding permeable beds. Young clay-mineral-bearing sediments at shallow burial depths are likely to be smectites (e.g., gumbo shale in the submerged Mississippi delta of Louisiana, US). Clay minerals in deeply buried shales become less hydrated, and their forms can be altered by higher temperatures and pressures. Low-salinity water is sometimes observed in reservoir-rock pores adjacent to the shales and where clay-mineral-expelled water cannot escape from the permeable bed. This formation may become overpressured and can cause severe problems during drilling if not predicted or detected in time to alter the drilling program.

Although shale formations are not usually of commercial importance, the measurement and evaluation of core and log data in partly shaly reservoirs presents many difficulties that are not present in clay-mineral-free (clean) formations. A large body of technical literature addresses shaly-formation analysis because the shale, to one degree or another, affects all log and core measurements. Because these shaly rocks are so variable, a single model usually cannot fully describe all of their behaviors.

Reservoirs with a fractional shale content, V_{sh}, are common, and the clay minerals/shales take several physical forms, including laminated, structural, and pore-filling. Laminated shales are thin detrital shale layers interbedded within a reservoir interval. Each represents short periods of deposition where the suspended finest sediments could settle out of the original sediment-rich river, lake, or seawater. Laminated shales may range from approximately one hundredth of a centimeter to 1 m thick. Shale deposits can be broken up and reworked after their original deposition and become “grains” in the same manner as quartz grains. Structural, or detrital, shale grains become a part of the grain composition of sandstone. As well as the clay minerals that are deposited directly as solids from lakes and marine environments, they also may be deposited from in-situ formation-water solutions that are rich in dissolved minerals. These clay minerals are called “authigenetic.” By this mechanism, the pores of sandstone may become partly filled with various clay minerals and other minerals. Of this type, illite, kaolinite, chlorite, and smectite clays are most common. Each can take several physical and chemical forms within a pore. Several generations of pore-filling clay minerals may be present, representing different periods of geological time when changes occurred in formation-water composition or depth of burial.

Evaluation of Shale Volume

Geological techniques, like XRD, are available to identify clay-mineral species and to quantify rock-component volumes in physical specimens. Such analyses can help calibrate the log-based methods for estimating V_{sh}, the bulk-volume fraction of shale. V_{sh} from the GR log is frequently used to determine nonpay.

Shale content can be estimated from well logs by many techniques, because shale affects the readings of most logs. The task for a particular field is to identify an evaluation technique that is reasonably accurate and as simple as possible. A method using a combination of the neutron and density logs is often applied for practical log analysis. Modern optimized simultaneous-equations log solutions attempt to identify individual clay species. Nevertheless, for petrophysical studies for field development, it is the GR log that is probably used most frequently to evaluate V_{sh}. The GR-log readings are normalized to reduce hole-size variations and mud effects and differences among the tools. Normalization is achieved by finding typical GR values in the 100% "clean" sand and 100% shale formations for each well. These different endpoint values in each well are then equalized. The GR values in between the sand and shale levels are scaled to give V_{sh} values. The scaling is often linear, but nonlinear alternatives are available, if appropriate. Occasionally the GR log is affected by radioactive components that are not shale, and these need to be identified and assigned a revised V_{sh}. Water-based drilling muds sometimes contain high concentrations of potassium salts, and these may also lead to GR interpretation problems, such as invasion of potassium salts into the near-wellbore region.

The uncertainty of V_{sh} log evaluations is moderate to high. At low values, less than 30% BV, the authors estimate that V_{sh} may typically be accurate to approximately ± 10% BV at one standard deviation (SD). At values greater than 30% BV, the uncertainty increases. The uncertainties affect nonpay bed-boundary evaluations; however, the V_{sh} uncertainty also seriously impacts the accuracy of the effective-porosity and S_w estimates when smectites, with their high clay-mineral adsorbed-water fraction, are present. If the clay species is different (illite or chlorite with little or no adsorbed water), then the V_{sh} uncertainty has a much-reduced impact on porosity.

Reservoir Zonation or Layering

An important conclusion from the geologists' technical studies is a definition of the extent to which the reservoir needs to be subdivided either vertically or areally. Besides all of the information developed by the geologists from the detailed core descriptions, the routine-core-analysis and SCAL data need to be analyzed for such effects. This is accomplished by preparing, for different possible layering within the reservoir interval, a variety of crossplots such as $\log_{10}$ (permeability) vs. porosity and OBM-core S_w vs. porosity or $\log_{10}$ (permeability) and comparing the data clouds and trends of those plots from one possible layer to the next. Significant differences should be expected between reservoir intervals with different depositional environments and differences in grain size and sorting. Areal variations may occur across a given vertical zone within the reservoir because of varying distances from the sediment source, differences in the depositional environment in various areas, or varying diagenetic effects.

For accurate petrophysical calculations, most large reservoirs will probably require a number of vertical subdivisions, usually termed zones or layers. Typical zones for a reservoir are 50 to 150 ft (15 to 45 m) thick. Areally, several square miles of reservoir can usually be included together; however, if the reservoir covers tens of square miles, it is likely to require some areal subdivision for accurate petrophysical calculations.

A second consideration is the amount of available data of various types. If a reservoir has a large database of log and core data from many wells, then the number of these subdivisions is not impacted by the quantity of data. However, if some types of data are very limited, then this consideration may control the degree of vertical and areal subdivision that can be used for various petrophysical calculations. The same degree of subdivision may not be required for net-pay calculations relative to porosity calculations or for porosity calculations relative to S_w calculations. In summary, more-accurate vertical and areal petrophysical calculations are made if the reservoir is appropriately subdivided. The figure below shows the vertical zonation used for the Prudhoe Bay field's Sadlerochit reservoir. The figures above are example plots based on real reservoir data that show how rock properties within the same field can vary from one reservoir to another and from one vertical portion of a reservoir interval to other parts.

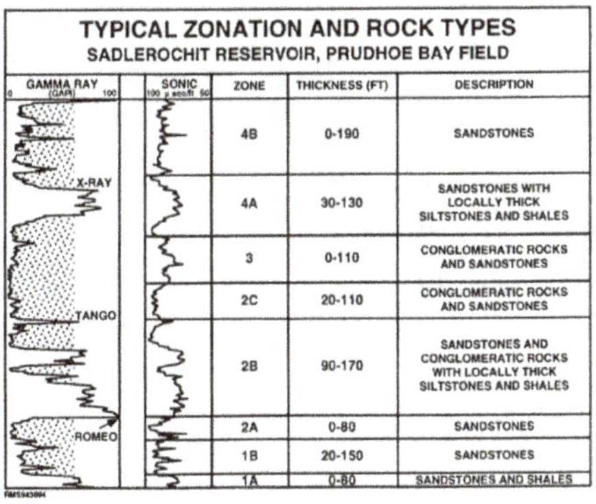

Figure: Reservoir zonation of the Sadlerochit reservoir, Prudhoe Bay field, Alaska, US.

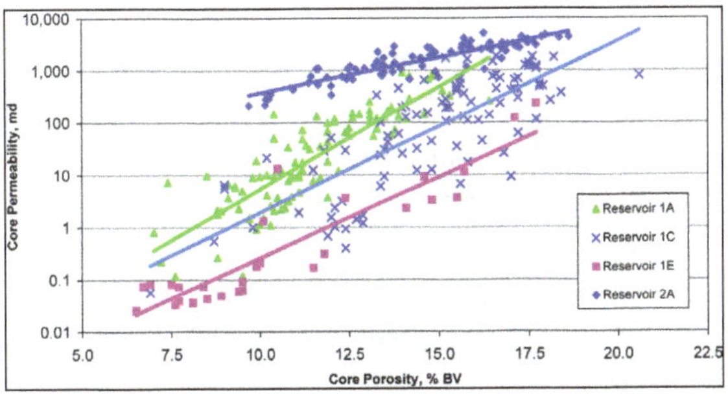

Figure: Typical reservoir permeability vs. porosity crossplot; data from an Asian gas field.

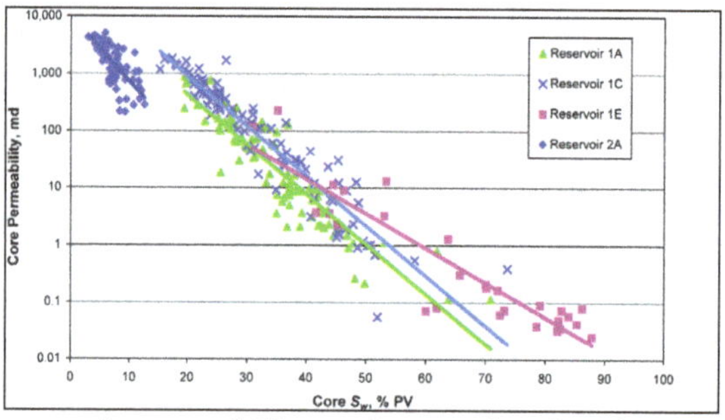

Figure: Typical reservoir permeability vs. S_w crossplot; data from an Asian gas field.

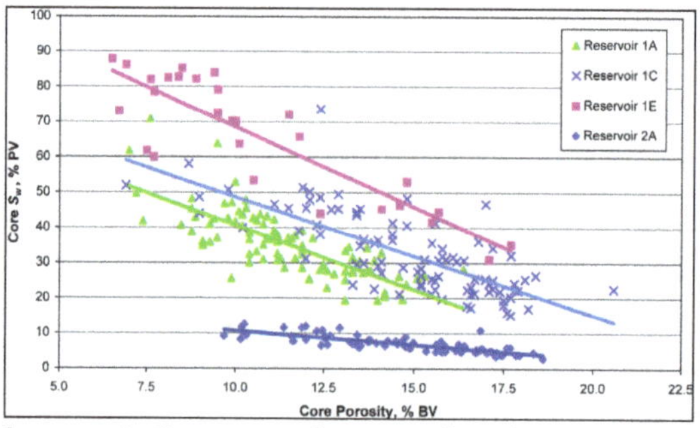

Figure: Typical reservoir S_w vs. porosity crossplot; data from an Asian gas field.

Lithostratigraphy

In lithostratigraphy rock units are considered in terms of the lithological characteristics of the strata and their relative stratigraphic positions. The relative stratigraphic positions of rock units can be determined by considering geometric and physical relationships that indicate which beds are older and which ones are younger. The units can be classified into a hierarchical system of members, formations and groups that provide a basis for categorising and describing rocks in lithostratigraphic terms.

Stratigraphic Relationships

Superposition

Provided the rocks are the right way up the beds higher in the stratigraphic sequence of deposits will be younger than the lower beds. This rule can be simply applied to a layer-cake stratigraphy but must be applied with care in circumstances where there is a significant depositional topography (e.g. fore-reef deposits may be lower than reef-crest rocks).

Unconformities

An unconformity is a break in sedimentation and where there is erosion of the underlying strata this provides a clear relationship in which the beds below the unconformity are clearly older than those above it. All rocks which lie above the unconformity, or a surface that can be correlated with it, must be younger than those below. In cases where strata have been deformed and partly eroded prior to deposition of the younger beds, an angular unconformity is formed. A disconformity marks a break in sedimentation and some erosion, but without any deformation of the underlying strata.

Cross-cutting Relationships

Any unit that has boundaries that cut across other strata must be younger than the rocks it cuts. This is most commonly seen with intrusive bodies such as batholiths on a larger scale and dykes on a smaller scale. This relationship is also seen in fissure fills, sedimentary dykes that form by younger sediments filling a crack or chasm in older rocks.

Included Fragments

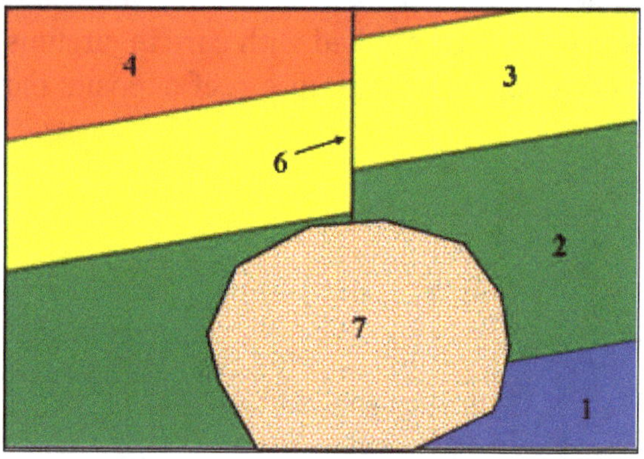

The fragments in a clastic rock must be made up of a rock that is older than the strata in which they are found. The same relationship holds true for igneous rocks that contain pieces of the surrounding country rock as xenoliths (literally 'foreign rocks'). This relationship can be useful in determining the age relationship between rock units that are some distance apart. Pebbles of a characteristic lithology can provide conclusive evidence that the source rock type was being eroded by the time a later unit was being deposited tens or hundreds of kilometres away.

Way-up Indicators in Sedimentary Rocks

The folding and faulting of strata during mountain building can rotate whole successions of beds (formed as horizontal or nearly horizontal layers) through any angle, resulting in beds that may be vertical or completely overturned. In any analysis of deformed strata, it is essential to know the direction of younging, that is, the direction through the layers towards younger rocks. The direction of younging can be determined by small-scale features that indicate the way-up of the beds or by using other stratigraphic techniques to determine the order of formation.

Lithostratigraphic Units

There is a hierarchical framework of terms used for lithostratigraphic units, and from largest to smallest these are: 'Supergroup', 'Group', 'Formation', 'Member' and 'Bed'. The basic unit of lithostratigraphic division of rocks is the formation, which is a body of material that can be identified by its lithological characteristics and by its stratigraphic position. It must be traceable laterally, that is, it must be mappable at the surface or in the subsurface. A formation should have some degree of lithological homogeneity and its defining characteristics may include mineralogical composition, texture, primary sedimentary structures and fossil content in addition to the lithological composition. Note that the material does not necessarily have to be lithified and that all the discussion of terminology and stratigraphic relationships applies equally to unconsolidated sediment. A formation is not defined in terms of its age either by isotopic dating or in terms of biostratigraphy. Information about the fossil content of a mapping unit is useful in the description of a formation but the detailed taxonomy of the fossils that may define the relative age in biostratigraphic terms does not form part of the definition of a lithostratigraphic unit. A formation may be, and often is, a diachronous unit, that is, a deposit with the same lithological properties that was formed at different times in different places. A formation may be divided into smaller units in order to provide more detail of the distribution of lithologies.

The term member is used for rock units that have limited lateral extent and are consistently related to a particular formation (or, rarely, more than one formation). An example would be a formation composed mainly of sandstone but which included beds of conglomerate in some parts of the area of outcrop. A number of members may be defined within a formation (or none at all) and the formation does not have to be completely subdivided in this way: some parts of a formation may not have a member status. Individual beds or sets of beds may be named if they are very distinctive by virtue of their lithology or fossil content. These beds may have economic significance or be useful in correlation because of their easily recognisable characteristics across an area. Where two or more formations are found associated with each other and share certain characteristics they are considered to form a group. Groups are commonly bound by unconformities which can be traced basin-wide. Unconformities that can be identified as major divisions in the stratigraphy over the area of a continent are sometimes considered to be the bounding surfaces of associations of two or more groups known as a supergroup.

Description of Lithostratigraphic Units

The formation is the fundamental lithostratigraphic unit and it is usual to follow a certain procedure in geological literature when describing a formation to ensure that most of the following issues are considered. Members and groups are usually described in a similar way.

Lithology and Characteristics

The field characteristics of the rock, for example, an oolitic grainstone, interbedded coarse siltstone and claystone, a basaltic lithic tuff, and so on form the first part of the description. Although a formation will normally consist mainly of one lithology, combinations of two or more lithologies will often constitute a formation as interbedded or interfingering units. Sedimentary structures (ripple cross-laminations, normal grading, etc.), petrography (often determined from thin-section analysis) and fossil content (both body and trace fossils) should also be noted.

Definition of Top and Base

These are the criteria that are used to distinguish beds of this unit from those of underlying and overlying units; this is most commonly a change in lithology from, say, calcareous mudstone to coral boundstone. Where the boundary is not a sharp change from one formation to another, but is gradational, an arbitrary boundary must be placed within the transition. As an example, if the lower formation consists of mainly mudstone with thin sandstone beds, and the upper is mainly sandstone with subordinate mudstone, the boundary may be placed at the point where sandstone first makes up more than 50% of beds. A common convention is for only the base of a unit to be defined at the type section: the top is taken as the defined position of the base of the overlying unit. This convention is used because at another location there may be beds at the top of the lower unit that are not present at the type locality: these can be simply added to the top without a need for redefining the formation boundaries.

Type Section

A type section is the location where the lithological characteristics are clear and, if possible, where the lower and upper boundaries of the formation can be seen. Sometimes it is necessary for a type section to be composite within a type area, with different sections described from different parts of the area. The type section will normally be presented as a graphic sedimentary log and this will form the strato type. It must be precisely located (grid reference and/or GPS location) to make it possible for any other geologist to visit the type section and see the boundaries and the lithological characteristics described.

Thickness and Extent

The thickness is measured in the type section, but variations in the thickness seen at other localities are also noted. The limits of the geographical area over which the unit is recognised should also be determined. There are no formal upper or lower limits to thickness and extent of rock units defined as a formation (or a member or group). The variability of rock types within an area will be the main constraint on the number and thickness of lithostratigraphic units that can be described and defined. Quality and quantity of exposure also play a role, as finer subdivision is possible in areas of good exposure.

Where the age for the formation can be determined by fossil content, radiometric dating or relationships with other rock units this may be included, but note that this does not form part of the definition of the formation. A formation would not be defined as, for example, 'rocks of Burdigalian age', because an interpretation of the fossil content or isotopic dating information is required to determine the age. Information about the facies and interpretation of the environment of deposition might be included but a formation should not be defined in terms of depositional environment, for example, 'lagoonal deposits', as this is an interpretation of the lithological characteristics.

Lithostratigraphic Nomenclature

It helps to avoid confusion if the definition and naming of stratigraphic units follows a set of rules. Formal codes have been set out in publications such as the 'North American

Stratigraphic Code' (North American Commission on Stratigraphic Nomenclature 1983) and the 'International Stratigraphic Guide'. A useful summary of stratigraphic methods, which is rather more user-friendly than the formal documents, is a handbook called 'Stratigraphical Procedure'. The name of the formation, group or member must be taken from a distinct and permanent geographical feature as close as possible to the type section. The lithology is often added to give a complete name such as the Kingston Limestone Formation, but it is not essential, or necessarily desirable if the lithological characteristics are varied. The choice of geographical name should be a feature or place marked on topographic maps such as a river, hill, town or village. The rules for naming members, groups and super groups are essentially the same as for formations, but note that it is not permissible to use a name that is already in use or to use the same name for two different ranks of lithostratigraphic unit. There are some exceptions to these rules of nomenclature that result from historical precedents, and it is less confusing to leave a well established name as it is rather than to dogmatically revise it. Revisions to stratigraphic nomenclature may become necessary when more detailed work is carried out or more information becomes available. New work in an area may allow a formation to be subdivided and the formation may then be elevated to the rank of group and members may become formations in their own right. For the sake of consistency the geographical name is retained when the rank of the unit is changed.

Lithodemic Units: Non-stratiform Rock Units

The concepts of division into stratigraphic units were developed for rock bodies that are stratiform, layered units, but many metamorphic, igneous plutonic and structurally deformed rocks are not stratiform and they do not follow the rules of superposition. Nonstratiform bodies of rock are called lithodemic units. The basic unit is the lithodeme and this is equivalent in rank to a formation and is also defined on lithological criteria. The word 'lithodeme' is itself rarely used in the name: the body of rock is normally referred to by its geographical name and lithology, such as the White River Granite or Black Hill Schist. An association of lithodemes that share lithological properties, such as a similar metamorphic grade, is referred to as a suite: the term complex is also used as the equivalent to a group for volcanic or tectonically deformed rocks.

Stratigraphy

The study of stratified rocks is known as stratigraphy. It is a branch of geology that deals with the correlation, interpretation, and description of stratified sediments and rocks that are found on or below the surface of the earth. Stratigraphy is a science that primarily deals with the study of layered sedimentary rocks (called strata) but may also include layered igneous rocks. Layered rocks may result from successive lava flows or from the formation of extrusive igneous rocks. Stratigraphy is an important field of modern archaeology. Since stratigraphy establishes that sedimentation occurs according to uniform principles, it is easier for archaeologists to draw up conclusions. When excavating, techniques used are based on the principles of stratigraphy.

Laws of Stratigraphy

The laws of stratigraphy were pioneered by Danish geologist, Nicholas Steno, who is also considered the father of stratigraphy. Steno established, through observation, that rock layers may not be chaotic as they seem. He concluded that rock layers (strata) are deposited with the upper layers being younger and the lower layers older. This, therefore, informs that rock layers indicate a chronological history of the earth and its past life. Stratigraphic records also preserve major geological events and details pertaining to orogenies (mountain-building), climatic fluctuation, and eustatically-driven sequences. As a result, Steno came up with three fundamental principles of stratigraphy, commonly known as Steno's laws.

Law of Original Horizontality

Due to gravitational settling, layers of sediment deposits are arranged in a horizontal manner. On tilted surfaces, the bed of sediments will tend to take a horizontal position while corresponding to the contours of the basin or depression. At the time of deposition, sedimentary layers assume a horizontal inclination. However, external forces may lead to deformation. The common types of deformation of horizontal layers include folding and faulting.

Law of Superposition

Oldest layers are found at the bottom and youngest layers at the top in undisturbed strata.

Law of Lateral Continuity

This principle establishes that horizontal layers stretch laterally until they thin to a negligible thickness at the edge of their basin of deposition.

Potential Pitfalls in Stratigraphy

To accurately study rock layers deposited on the earth, the strata must be undisturbed and unadulterated. There is plenty of room for error when undertaking this kind of study. There are many challenges and pitfalls that the geologist can encounter in the study of sedimentary deposits.

First, layers can be less than an inch deep and some can be many feet deep. It is on the geologist to use accurate judgment to determine whether erosion and other geological forces have affected the stratum. In a case where erosion has occurred, it may be difficult to come up with an accurate stratigraphic record due to disturbance and unconformity. As such the study may be rendered inaccurate.

The stratigraphic record may be difficult to read in the case where an earthquake disturbance has taken place. Under such circumstances, part of the earth's crust may shift over an adjacent section. The geologist will find it difficult to establish neat lines that define the stratum. Additionally, the change in thickness of stratum extends towards the edges where it merges with other deposits.

Geologists also face a major pitfall when it comes to establishing the age of the strata using radiometric dating. The problem arises when attempting to find the absolute age of materials from which the strata are formed. A sample taken from the stratum may indicate the age and period of the parent rock instead of the sample itself. For example, grains of sand that constitute a sandstone are older than the larger piece of sandstone. Radiometric reading will indicate the age of the grain sands as opposed to the unit of sandstone. This provides a unique challenge to overcome.

Importance of Stratigraphy

Stratigraphy deals with different periods of geological past by revealing their details from the beginning to the present period. It is, therefore, an important field in explain various environmental effects and how they shaped our present.

Stratigraphy enables us to understand the details of geography, climate, glaciations, orogeny, epeirogeny, and how the various factors affected the evolution and migration of plants and animals.

Through stratigraphic analyses, scientists and geologists are provided with valuable clues about the age of the earth and its history. These findings are used in a myriad of other scientific fields and disciplines ranging from city planners, prospecting geologists, petrologists, and engineers.

Archaeologists and paleontologist rely on stratigraphy to provide context to their findings during the excavation process. They can accurately determine the history of various animals dating back to historical and pre-historical periods.

Permissions

All chapters in this book are published with permission under the Creative Commons Attribution Share Alike License or equivalent. Every chapter published in this book has been scrutinized by our experts. Their significance has been extensively debated. The topics covered herein carry significant information for a comprehensive understanding. They may even be implemented as practical applications or may be referred to as a beginning point for further studies.

We would like to thank the editorial team for lending their expertise to make the book truly unique. They have played a crucial role in the development of this book. Without their invaluable contributions this book wouldn't have been possible. They have made vital efforts to compile up to date information on the varied aspects of this subject to make this book a valuable addition to the collection of many professionals and students.

This book was conceptualized with the vision of imparting up-to-date and integrated information in this field. To ensure the same, a matchless editorial board was set up. Every individual on the board went through rigorous rounds of assessment to prove their worth. After which they invested a large part of their time researching and compiling the most relevant data for our readers.

The editorial board has been involved in producing this book since its inception. They have spent rigorous hours researching and exploring the diverse topics which have resulted in the successful publishing of this book. They have passed on their knowledge of decades through this book. To expedite this challenging task, the publisher supported the team at every step. A small team of assistant editors was also appointed to further simplify the editing procedure and attain best results for the readers.

Apart from the editorial board, the designing team has also invested a significant amount of their time in understanding the subject and creating the most relevant covers. They scrutinized every image to scout for the most suitable representation of the subject and create an appropriate cover for the book.

The publishing team has been an ardent support to the editorial, designing and production team. Their endless efforts to recruit the best for this project, has resulted in the accomplishment of this book. They are a veteran in the field of academics and their pool of knowledge is as vast as their experience in printing. Their expertise and guidance has proved useful at every step. Their uncompromising quality standards have made this book an exceptional effort. Their encouragement from time to time has been an inspiration for everyone.

The publisher and the editorial board hope that this book will prove to be a valuable piece of knowledge for students, practitioners and scholars across the globe.

Index